AF445178

After suffering an attack against his exemplary and impeccable professional career, Engineer Israel Laisequilla offers us in this work a detailed and clearly written guide that allows us to delve into the world of industry in the unique manner that only the so-called "most controversial engineer" can achieve.

INSTRUCTIONS

The information presented here takes into account the access and/or availability of current information and technologies. In case you require more information, formats, and/or examples, your inquiry may increase its reliability thanks to the criteria developed with the book. Any additional information, formulas, and/or videos may be requested using your preferred artificial assistant.

The Bible of the Industrial Engineer

Engineering and Methods

ENGR'S WORKSHOP

I. LAISEQUILLA

Copyright © 2024 Israel Laisequilla

Revision 1

Author's introduction added, July 2024.

ii

To the industrial engineers

CONTENT

INDUSTRIAL METHODS

ACKNOWLEDGMENTS

I would like to express my deep gratitude to those who have made the creation of this book possible, a work that arises from the fusion of two previous works and represents the culmination of years of dedication and passion for industrial engineering.

First and foremost, my eternal gratitude to my beloved family. Their constant support, understanding, and patience have been the foundation upon which this project has been built. Thank you for being my endless source of inspiration.

I extend special thanks to my editor, whose experience, vision, and dedication have been crucial in shaping this book. Their expert guidance and commitment to excellence have elevated this project to new heights, and I am profoundly grateful for their collaboration.

To the valuable readers of my previous works and those who are now venturing into exploring The bible of Industrial Engineer - Engineering and Methods, I extend my sincerest thanks. Your interest and support are the driving force behind my work, and I hope you find valuable knowledge and enriching perspectives within these pages.

This book is the result of a collective effort, and to each person who has contributed in any way, whether with words of encouragement, constructive feedback, or collaborative efforts, I am grateful.

May this compendium be a source of inspiration and knowledge for the curious and passionate minds in the field of industrial engineering.

engr's Workshop

INTRODUCTION TO INDUSTRIAL ENGINEERING

Industrial engineering is a discipline that focuses on the design, improvement, and management of systems and processes for the production of goods and services. This discipline is based on the principles of engineering, science, and management, and encompasses a wide range of fields including operations management, logistics, production, ergonomics, occupational safety, automation, and control, among others.

The main objective of industrial engineering is to optimize processes and systems to improve efficiency, reduce costs, increase quality, and ensure the safety and well-being of workers and consumers. To achieve these goals, industrial engineers use advanced tools and techniques for analysis, design, and management, as well as a wide variety of technologies and methodologies.

The role of industrial engineering in society is crucial as it enables efficient and effective production of goods and services, which in turn contributes to economic growth and overall well-being. Industrial engineering is used in a wide range of sectors including manufacturing, energy, healthcare, transportation, communications, and financial services.

The evolution of industrial engineering

Industrial engineering has evolved over the years, with its origins dating back to the Industrial Revolution in the 18th century. At that time, the manufacturing industry was in a period of growth and expansion, and the need to increase efficiency and productivity was evident. The invention of the steam engine and other technological advancements allowed for the creation of factories and mass production, which led to a greater demand for engineers and management experts.

Over time, industrial engineering has expanded to encompass a wide range of fields, from production management to logistics and supply chain management. The discipline has also evolved to include greater attention to sustainability and social responsibility as environmental and social issues have become more prominent in society.

The Fundamentals of Industrial Engineering

Industrial engineers use a variety of tools and techniques to optimize processes and systems. Some of these tools include operations research, statistics, simulation, linear programming, optimization, and systems modeling. Industrial engineers also use a variety of management techniques, such as project management, quality management, change management, and supply chain management.

The work of an industrial engineer involves a combination of technical and management skills, which means that industrial engineers must have a strong background in mathematics, physical and social sciences, as well as management and leadership. Additionally, industrial engineers must be able to work in teams, communicate effectively, and have critical thinking skills to address and solve complex problems.

The field of industrial engineering also involves a deep understanding of ergonomics and occupational safety, as industrial engineers work closely with workers in production processes and must ensure that workers are safe and comfortable while performing their tasks.

Another important aspect of industrial engineering is automation and control, which refers to the implementation of systems and technologies to improve efficiency and quality in production processes. Industrial engineers work with robots, control systems, and advanced technologies to optimize processes and reduce costs.

Applications of Industrial Engineering

Industrial engineering is used in a wide variety of industries and sectors. In the manufacturing industry, industrial engineers work on mass production, automation, robotics, and quality improvement. In the healthcare industry, industrial engineers may work on designing and implementing efficient and effective healthcare systems. In the transportation industry, industrial engineers may work on supply chain management and logistics to ensure timely and efficient delivery of goods and services.

In the financial services sector, industrial engineers may work on process and system management to ensure efficiency and security in financial transactions. In the energy sector, industrial engineers may work on improving energy efficiency and implementing sustainable technologies.

In addition, industrial engineering also has applications in the public sector, where industrial engineers may work on improving efficiency and quality in government processes and the provision of public services.

Challenges and Opportunities in Industrial Engineering

Industrial engineering faces a series of challenges and opportunities today. One of the biggest challenges is the need to address environmental and social issues, such as sustainability and social responsibility. Industrial engineers must find ways to reduce the environmental impact of production processes and ensure that business practices are socially responsible.

Another major challenge is the need to adapt to technological advances and the digitalization of production processes. Industrial engineers must stay up-to-date on the latest technologies and tools to ensure that processes and systems are efficient and effective.

In addition, globalization and international competition have increased the need to improve efficiency and quality in production processes. Industrial engineers must find ways to compete in a globalized market and ensure that their processes are efficient and profitable.

Despite these challenges, industrial engineering also offers many exciting opportunities for those seeking a career in this field. The growing demand for industrial engineers means that there is a wide range of employment opportunities in a variety of industries and sectors. In addition, industrial engineering is a constantly evolving discipline, which means that there are many opportunities for professional growth and innovation in the field.

In summary, industrial engineering is a fascinating discipline that combines a wide range of skills and knowledge to improve efficiency and quality in production processes. From supply chain management to quality improvement, automation, and the implementation of advanced technologies, industrial engineers play a critical role in the success of a variety of industries and sectors. With a growing demand for professionals in this field, industrial engineering offers an exciting range of employment and professional growth opportunities.

FUNDAMENTALS OF OPERATIONS MANAGEMENT

Operations management is a key discipline in industrial engineering, as it focuses on the planning, coordination, and control of production and service processes. In this chapter, we will explore the fundamentals of operations management and how they are applied in practice.

Basic concepts of operations management

Operations management focuses on the efficiency and effectiveness of production and service processes. Its goal is to optimize the use of resources, improve the quality of the product or service, and increase customer satisfaction. To achieve this, specific tools and techniques are used, such as production planning, inventory management, production scheduling, quality management, and supply chain management.

Production planning

Production planning is a key process in operations management. It involves defining production goals, determining the necessary resources, and scheduling the activities needed to produce the product or service. Production planning can be divided into three levels: long-term planning, medium-term planning, and short-term planning.

In long-term planning, long-term production goals are established, necessary resources are determined, and investment plans are defined. In medium-term planning, medium-term production plans are defined, activities needed to meet long-term goals are scheduled, and required capacity is determined. In short-term planning, short-term production plans are defined, activities needed to meet medium-term goals are scheduled, and required capacity is determined.

Inventory management

Inventory management refers to the control and monitoring of the quantity and value of products or materials stored in a warehouse. This management is important to ensure the availability of necessary products and materials for production, but also to avoid excessive inventory accumulation that may generate unnecessary costs. Specific techniques such as reorder point, safety inventory, and ABC analysis are used for inventory management.

Production scheduling

Production scheduling is a process that involves the allocation of resources and scheduling of necessary activities to produce the product or service. Specific techniques such as Gantt chart, PERT method, and linear programming are used for production scheduling. These techniques enable effective and efficient production activity scheduling.

Quality management

Quality management refers to a set of activities performed to ensure that a product or service meets customer requirements and expectations. This management is important to ensure customer satisfaction and to avoid unnecessary costs associated with customer dissatisfaction. Specific techniques such as statistical quality control, acceptance sampling, and cause and effect analysis are used for quality management.

Supply chain management

Supply chain management refers to the coordination and management of processes from the acquisition of necessary materials and components for production to the delivery of the final product or service to the end customer. This management is important to ensure the availability of necessary materials and components, as well as to ensure timely and efficient delivery of the product or service. Specific techniques such as demand planning, supplier management, and warehouse management are used for supply chain management.

Tools and Techniques for Operations Management

In addition to the basic concepts of operations management, there are several tools and techniques used to improve the efficiency and effectiveness of production and service processes.

Lean Manufacturing

Lean manufacturing is a methodology that focuses on the elimination of waste in production processes. It is based on five principles: identifying value, mapping the value stream, creating continuous flow, establishing a pull system, and striving for perfection. By implementing lean manufacturing, companies can improve efficiency and reduce costs.

Six Sigma

Six Sigma is a methodology that focuses on reducing variability in production processes. It is based on measuring, analyzing, and improving processes to achieve significant reduction in defects. By implementing Six Sigma, companies can improve the quality of their products or services and reduce costs.

Theory of Constraints

The theory of constraints is a methodology that focuses on identifying and eliminating limitations in production processes. It is based on identifying the most critical constraint and implementing measures to eliminate or reduce it. By implementing the theory of constraints, companies can improve the efficiency of their processes and reduce production times.

Just-in-Time

Just-in-Time is a methodology that focuses on eliminating unnecessary inventory. It is based on the production and delivery of products or materials just when they are needed. By implementing Just-in-Time, companies can reduce the costs associated with inventory management and improve the efficiency of their processes.

Conclusion

In summary, operations management is a key discipline in industrial engineering, focusing on the planning, coordination, and control of production and service processes. Operations management involves production planning, inventory management, production scheduling, quality management, and supply chain management. Additionally, there are specific tools and techniques, such as lean manufacturing, Six Sigma, the theory of constraints, and Just-in-Time, used to improve the efficiency and effectiveness of production and service processes.

DESIGN OF INDUSTRIAL SYSTEMS AND PROCESSES

The design of industrial systems and processes is a fundamental part of industrial engineering. Adequate design of systems and processes can improve efficiency, productivity, quality, and safety in industrial environments. In this chapter, we will explore the fundamentals of industrial systems and processes design, as well as some of the tools and techniques used in this process.

Systems Design

Systems design involves creating a set of interconnected components and subsystems that work together to achieve a specific objective. Systems design can be applied to a wide variety of industrial applications, including manufacturing, transportation, energy, health, and safety.

In system design, it is important to consider the needs and expectations of the end-user, as well as technical and economic constraints. In addition, system design should be able to support future changes and adaptations.

The process of system design involves several stages, which may include defining requirements, identifying design alternatives, evaluating alternatives, selecting the preferred design, and implementing and validating the design.

Process Design

Process design involves creating a set of interconnected activities that convert raw materials and energy into products or services. Proper process design can improve efficiency, quality, safety, and sustainability in industrial environments.

In process design, it is important to consider the needs and expectations of the customer, as well as technical and economic constraints. In addition, process design should be able to support future changes and adaptations.

The process of process design involves several stages, which may include defining requirements, identifying design alternatives, evaluating alternatives, selecting the preferred design, and implementing and validating the design.

Tools and Techniques for Designing Systems and Processes

There are several tools and techniques that can be used in the design of industrial systems and processes. Some of the most common tools and techniques include:

Flowcharts: Flowcharts are tools used to visualize and analyze processes. Flowcharts can help identify problem areas in processes and design solutions to improve the efficiency and effectiveness of processes.

Risk Analysis: Risk analysis is a technique used to identify and evaluate the risks associated with systems and processes. Risk analysis can help companies design solutions to mitigate risks and improve the safety of systems and processes.

Simulation: Simulation is a technique used to model and analyze systems and processes. Simulation can help companies identify problem areas in systems and processes and design solutions to improve the efficiency and effectiveness of systems and processes.

Design for Manufacturability: Design for manufacturability is a technique used for designing industrial systems and processes. In the industry, the design of systems and processes is one of the most important tasks to ensure the quality and efficiency of production. In this chapter, we will explore the importance of designing effective systems and processes in industrial engineering, as well as the key principles that engineers must consider when designing effective systems and processes.

Introduction to Designing Systems and Processes

Designing industrial systems and processes is a key process that involves creating and developing effective systems and processes for the production and manufacturing of products. In general terms, designing industrial systems and processes involves a wide range of activities, from initial planning and design of systems and processes to the implementation and optimization of existing systems and processes.

In industrial engineering, designing systems and processes is a critical task to ensure the quality, efficiency, and profitability of production. Industrial engineers are dedicated to creating optimized systems and processes to maximize productivity, minimize costs, and ensure product quality. By designing effective systems and processes, industrial engineers can improve the competitiveness and profitability of the company.

Key principles of industrial systems and processes design

The design of industrial systems and processes involves a wide range of activities, but there are some key principles that industrial engineers must keep in mind when designing effective systems and processes. These principles include:

Defining the objectives and requirements of the system or process: Before starting the design of a system or process, it is important to clearly define its objectives and requirements. Industrial engineers must take into account the company's objectives, customer needs, and resource limitations when designing a system or process.

Identifying key processes: Effective design of a system or process requires the identification and analysis of the key processes involved. Industrial engineers must understand how processes are performed and how they relate to each other to design effective systems and processes.

Designing for quality: Product quality is a critical objective in the design of industrial systems and processes. Industrial engineers must design systems and processes that minimize variability and maximize product quality.

Designing for efficiency: Efficiency is another important objective in the design of industrial systems and processes. Industrial engineers must design systems and processes that minimize waste of materials, energy, and time.

Designing for flexibility: Industrial systems and processes must be designed to be flexible and adaptable to changes in customer needs or production. Industrial engineers must anticipate possible changes and design systems and processes that can adapt to them.

Continuously evaluating and improving: The design of industrial systems and processes is a continuous process that requires constant evaluation and improvement. Industrial engineers must regularly analyze existing systems and processes to identify opportunities for improvement. In addition to process flow analysis, another tool used in the design of industrial systems and processes is simulation. Simulation allows industrial engineers to create a mathematical model of a system or process and use it to predict its behavior under different conditions. This is especially useful when dealing with complex systems or processes, as it can be difficult to predict how they will perform in real life.

Simulation is often performed using specialized software that can model complex systems and processes. These models can be very detailed and include multiple variables and factors. Industrial engineers can use simulation to test different scenarios and see how they affect system or process performance. They can also use simulation to identify bottlenecks and other performance problems that may not be evident otherwise.

The design of systems and processes also involves selecting appropriate equipment and technologies. Industrial engineers must understand the capabilities and limitations of different types of equipment and technologies in order to select the most suitable one for each task. This may include selecting machinery, tools, materials, and other resources.

Once the appropriate equipment and technology have been selected, industrial engineers can begin designing production processes. This includes creating detailed flowcharts that show the movement of materials and workers through the process. These flowcharts can also help industrial engineers identify areas where efficiency can be improved and costs reduced.

In addition to planning and design, the process of designing industrial systems and processes also includes implementation and commissioning. During this phase, industrial engineers work with plant workers and managers to implement the new systems and processes. They may also provide training and support to ensure that all employees understand the new system or process and are comfortable using it.

Finally, industrial engineers must monitor and measure the performance of the new system or process to ensure that it is functioning as intended. This involves measuring and analyzing key metrics such as performance, efficiency, and costs. Industrial engineers may also use feedback from employees and managers to identify areas where further improvements can be made.

In summary, the design of industrial systems and processes is a fundamental aspect of industrial engineering. It involves the planning, design, implementation, and measurement of the performance of systems and processes used in industrial production. Industrial engineers use a variety of tools and techniques, including process flow analysis, simulation, and equipment and technology selection, to create efficient and cost-effective systems and processes that help companies achieve their production objectives.

QUALITY CONTROL AND CONTINUOUS IMPROVEMENT

Quality control and continuous improvement are two critical concepts in any organization that seeks to improve its processes and offer high-quality products or services to its customers. Quality control is a process that seeks to ensure that products or services meet customer requirements and specifications and are delivered consistently. Continuous improvement, on the other hand, is a process that seeks to continuously improve the organization's processes and products over time. In this chapter, we will explore the concepts of quality control and continuous improvement in depth, analyzing their key components and their impact on the organization.

Quality Control

Quality control is a process that focuses on ensuring that products or services offered meet customer requirements and specifications. This process is based on identifying problems or deviations in processes and products, implementing measures to correct these problems, and measuring the effectiveness of these measures. Quality control focuses on ensuring that products or services meet the quality specifications established by the organization, industry, or customers.

Quality control is divided into two main types: internal quality control and external quality control. Internal quality control focuses on ensuring that products or services meet the organization's internal requirements and specifications. External quality control, on the other hand, focuses on ensuring that products or services meet the requirements and specifications of customers or external regulations.

Components of Quality Control

There are several key components in quality control. These include planning, process control, quality evaluation, continuous improvement, and a customer-focused approach.

Planning

Planning is a key component of quality control. In this process, quality objectives are established, critical processes and products are identified, quality specifications are set, and quality control plans are developed. Planning also includes identifying risks and implementing measures to mitigate them.

Process Control

Process control is a key component of quality control. This process focuses on ensuring that the organization's processes are running optimally and meet established quality requirements and specifications. Process control includes identifying critical process points, implementing measures to control these points, and measuring the effectiveness of these measures.

Quality Evaluation

Quality evaluation is another key component of quality control. This process focuses on measuring the quality of the products or services offered. Quality evaluation includes measuring the product or service's conformity to established quality specifications, identifying problems or deviations, and implementing measures to correct these problems.

Continuous Improvement

Continuous improvement is a key component of quality control. This process focuses on identifying areas for improvement in the organization's processes and products and implementing measures to continuously improve these processes and products. Continuous improvement is based on Deming's continuous improvement cycle, which consists of four steps: plan, do, check, and act. In the first step, improvements to be implemented are planned, in the second step, planned improvements are implemented, in the third step, the effectiveness of implemented improvements is checked, and in the fourth step, action is taken based on the results obtained to continue improving the process.

Customer Focus

Customer focus is another key component of quality control. This process focuses on understanding customer needs and expectations and ensuring that the products or services offered meet these needs and expectations. Customer focus includes customer feedback and implementing measures to improve customer satisfaction.

Quality Control Tools

There are several quality control tools used to improve processes and ensure the quality of products or services. Some of the most common tools include:

Pareto Chart: It is a tool used to identify critical problems in a process and prioritize them based on their impact on the quality of the product or service.

Ishikawa Diagram: Also known as a fishbone diagram, it is a tool used to identify the causes of a problem and understand how these causes are related.

Control Charts: These are tools used to monitor the quality of a process over time and detect deviations that may indicate problems in the process.

Statistical Sampling: It is a tool used to measure the quality of a process through the collection and analysis of a sample of products or services.

Continuous Improvement

Continuous improvement is a critical process for any organization that seeks to ensure the quality of its products or services and continuously improve its processes. Continuous improvement focuses on identifying areas for improvement and implementing measures to continuously improve the organization's processes and products.

Deming's Continuous Improvement Cycle is a tool used to implement continuous improvement in an organization. This cycle consists of four steps: plan, do, check, and act.

Plan: In this step, areas for improvement are identified and plans are developed to implement improvements in these areas.

Do: In this step, planned improvements are implemented.

Check: In this step, the effectiveness of the implemented improvements is verified through measurement and analysis of the results.

Act: In this step, actions are taken based on the results obtained, and the process continues to be improved.

Continuous improvement is a process that requires the participation and commitment of all members of the organization. It is important to establish a culture of continuous improvement in the organization and encourage the participation of all members in identifying areas for improvement and implementing improvements.

Benefits of Quality Control and Continuous Improvement

Quality control and continuous improvement have many benefits for an organization. Some of these benefits include:

Improvement of the quality of the products or services offered.

Reduction of costs associated with the production and delivery of products or services.

Improvement of the efficiency and effectiveness of processes.

Improvement of customer satisfaction.

Increased customer loyalty and retention.

Reduction of the product return rate or service cancellations.

Improvement of the organization's reputation.

Increase of the organization's competitiveness.

Improvement of the work environment by involving employees in the process of continuous improvement.

Overall, quality control and continuous improvement are fundamental to the long-term success of an organization. By implementing these processes, an organization can ensure that its products or services meet the expectations of its customers and continuously improve its processes to remain competitive in a constantly changing market.

Application Example of Quality Control and Continuous Improvement

To illustrate how quality control and continuous improvement are applied in practice, the example of a car manufacturing company can be used.

Planning: In this phase, areas for improvement in the automobile manufacturing process are identified. In this case, the painting process can be identified as a possible area for improvement, as some inconsistencies in the color and quality of the paint have been identified in some of the manufactured vehicles.

Doing: In this phase, the planned improvements for the painting process are implemented. Improvements can be made to the paint mixing and application process to ensure that it is applied uniformly and meets the required color and quality standards.

Verification: In this phase, the effectiveness of the implemented improvements is verified through measurement and analysis of the results. In this case, the quality of the paint applied to the manufactured vehicles can be measured and compared with previous results to determine if the improvements have had a positive impact.

Acting: In this phase, action is taken based on the results obtained, and the process continues to be improved. If the results are positive, the improvements in the painting process can be implemented in all future manufactured vehicles. If the results are unsatisfactory, other areas for improvement can be identified, and the continuous improvement process can start again.

Conclusions

In conclusion, quality control and continuous improvement are critical processes for any organization seeking to ensure the quality of its products or services and continually improve its processes. These processes can help reduce costs, increase efficiency and effectiveness, improve customer satisfaction, and increase the organization's competitiveness.

Customer focus is fundamental to the success of quality control and continuous improvement as it allows organizations to understand customer needs and expectations and ensure that their products or services meet these needs and expectations.

Quality control tools are valuable for identifying areas for improvement and ensuring the quality of products or services. However, it is important to remember that these tools should be used in conjunction with a holistic approach to continuous improvement to achieve optimal results.

In summary, quality control and continuous improvement are dynamic and ongoing processes that require the participation and commitment of all members of the organization. By implementing these processes effectively, an organization can ensure the quality of its products or services, meet customer needs and expectations, and remain competitive in a constantly changing market.

METHODS OF ANALYSIS AND OPTIMIZATION

The chapter on methods of analysis and optimization is one of the fundamental pillars in decision-making and problem-solving in a wide variety of fields. In this chapter, techniques and tools are explored for analyzing and optimizing systems, processes, projects, among others. On this occasion, we will address the main concepts and techniques used in the optimization and analysis of systems and processes, as well as some practical examples of their application.

Fundamental concepts

Before delving into the techniques and tools for analyzing and optimizing systems and processes, it is important to define some fundamental concepts. Firstly, optimization refers to the process of searching for the best possible solution within a set of possible options. Secondly, analysis refers to the process of understanding how a system or process works, identifying its strengths and weaknesses, and detecting opportunities for improvement.

Another fundamental concept in the optimization and analysis of systems and processes is that of the objective function. The objective function is a quantitative measure of what is desired to be optimized or maximized. For example, in a production process, the objective function may be to maximize production and minimize costs. In a transportation system, the objective function may be to minimize travel time or maximize fuel efficiency.

Optimization methods

There are numerous methods and techniques for optimizing systems and processes, each with its own advantages and disadvantages. Below are some of the most common methods.

Trial and error method

The trial and error method is one of the simplest methods for optimizing a system or process. In this method, different configurations or adjustments are tested until the best solution is found. Although it is an intuitive and easy-to-implement method, it can be inefficient and time-consuming.

Gradient Descent Method

The Gradient Descent Method is an iterative method for finding the minimum of an objective function. In this method, a random point is selected, and the direction and magnitude of the gradient of the objective function are calculated. Then, the point moves in the opposite direction to the gradient, in which the objective function decreases. This process is repeated until a local minimum is reached.

Random Search Method

The Random Search Method consists of randomly generating a solution and evaluating its objective function value. Then, new random solutions are generated and evaluated, and the process is repeated until the best possible solution is found. Although this method is simple and easy to implement, it can be very inefficient and require many evaluations of the objective function.

Linear Programming Method

Linear programming is a mathematical method for optimizing a linear objective function subject to linear constraints. In this method, decision variables are defined, and constraints are established on them. Then, the optimal solution that maximizes or minimizes the objective function subject to the constraints is sought. Linear programming is a powerful and efficient method for solving linear optimization problems.

Integer Programming Method

Integer programming is an extension of linear programming, in which the decision variables are restricted to integer values. This makes the problem more complex and difficult to solve, but it allows modeling a wide variety of real-world problems, such as resource allocation and production scheduling.

Simulation Method

The simulation method involves creating a mathematical model of the system or process that needs to be optimized, and then simulating its behavior to evaluate different scenarios and configurations. This method allows for analyzing the impact of different decisions on the system or process without having to conduct real-world experiments. Simulation is a powerful tool for the analysis and optimization of complex systems and processes.

Sensitivity Analysis Method

The sensitivity analysis method is used to evaluate how the objective function changes when the model parameters are altered. For example, if the objective function is the production of a factory, the parameters could be the costs of materials, processing times, etc. Sensitivity analysis helps identify which parameters have a significant impact on the objective function and which are less important.

Practical Example

To illustrate the use of some of these optimization and analysis methods, let's consider the following example. A transportation company wants to optimize its truck fleet to minimize fuel costs and maximize transport efficiency. The company has a fleet of trucks of different sizes and capacities and needs to decide which trucks to assign to which routes to minimize costs.

To solve this problem, we can use a linear programming model to optimally assign the trucks to the routes. The model can include decision variables for truck assignment to routes, as well as constraints on truck capacity and route demand. The objective function can be a combination of fuel costs and transport efficiency, weighted by their relative importance.

Once the model has been constructed, a linear programming software can be used to find the optimal solution. Sensitivity analysis can then be performed to evaluate how the optimal solution changes when the model parameters, such as fuel prices or route demand, are altered.

Another approach to optimize the truck fleet is to use the simulation method. In this approach, a mathematical model of the transportation system is constructed and its behavior is simulated to evaluate different scenarios and configurations. For example, the impact of changing the truck assignment to routes or adding new trucks to the fleet can be simulated.

Conclusions

The chapter on Methods of Analysis and Optimization is essential for problem-solving and decision-making in a wide variety of fields. In this chapter, we have reviewed the fundamental concepts of optimization and analysis, as well as the most common methods used in practice. Additionally, we have presented a practical example of the application of these methods in the context of optimizing a fleet of trucks.

It is important to note that although there are different methods of analysis and optimization, it is crucial to select the appropriate method for each specific problem. For example, linear programming is suitable for problems that can be modeled as linear equations, while integer programming is necessary when decision variables must be restricted to integer values. Similarly, the simulation method is suitable for complex systems and processes that cannot be easily modeled using mathematical equations.

Furthermore, it is important to remember that the mathematical models used in optimization and analysis are simplifications of reality, and there will always be limitations and assumptions in the model. Therefore, it is important to validate the results obtained through simulation or sensitivity analysis by comparison with real-world data.

In summary, the chapter on Methods of Analysis and Optimization is fundamental for problem-solving and decision-making in a wide variety of fields. The methods presented in this chapter, such as linear programming, integer programming, the simulation method, and sensitivity analysis, are powerful tools for the analysis and optimization of complex systems and processes. However, it is important to select the appropriate method for each specific problem, validate the results, and consider the limitations and assumptions in the model used.

SIMULATION MODELS AND DECISION-MAKING

Simulation is a widely used tool in decision-making in different fields, from engineering to finance and health. It allows modeling complex systems and predicting their behavior in different situations. Decision-making based on simulation involves the use of mathematical models to simulate different scenarios and evaluate the impact of each one. In this chapter, simulation models and their application in decision-making will be discussed.

Types of Simulation Models

There are several types of simulation models used in different fields. Some of the most common types are as follows:

Discrete event models: This type of model is used to simulate systems in which events occur at specific and discrete times. Examples include queues in supermarkets or airports, production processes, and transportation systems.

Dynamic system models: This type of model is used to simulate systems in which variables change continuously over time. Examples include weather systems, economic systems, and biological systems.

Agent-based simulation models: This type of model is used to simulate systems in which individual agents have autonomous behaviors and can interact with each other. Examples include traffic simulations, market simulations, and animal behavior simulations.

Each type of model has its own advantages and disadvantages, and the choice of model type depends on the system being simulated and the objectives of the simulation.

Stages of Simulation

The simulation process consists of several stages, including problem formulation, model construction, model validation, simulation execution, and results analysis.

Problem Formulation: In this stage, the problem to be simulated is defined and simulation objectives are established. The variables that influence the system are also identified and model parameters are defined.

Model Construction: In this stage, the mathematical model representing the system to be simulated is constructed. The type of model to be used is selected and equations describing the system's behavior are established.

Model Validation: In this stage, it is verified that the model is valid and accurately reflects the behavior of the real system. Simulation results are compared with historical or experimental data to evaluate the accuracy of the model.

Simulation Execution: In this stage, the simulation is executed using the model constructed and parameters defined in the previous stages.

Results Analysis: In this stage, simulation results are analyzed to evaluate the impact of different scenarios and make decisions based on the results.

Applications of Simulation in Decision Making

Simulation is used in a wide variety of fields for decision making. Some of the most common applications include:

Engineering Simulation: Simulation is used in engineering to model complex systems and predict their behavior in different situations. For example, it can be used to simulate fluid flow in a pipeline system, evaluate the performance of a production system, or predict the behavior of a structure during an earthquake.

Simulation in finance: Simulation is used in finance to model different economic scenarios and evaluate their impact on an investment portfolio or a company. For example, it can be used to evaluate investment risk, simulate the behavior of stock prices, or evaluate the impact of different investment strategies.

Simulation in healthcare: Simulation is used in healthcare to model complex biological systems and predict their behavior in different situations. For example, it can be used to simulate the behavior of a virus in a population, evaluate the impact of different treatments on patients, or simulate the behavior of an immune system.

Simulation in logistics: Simulation is used in logistics to model transportation systems and predict their behavior in different situations. For example, it can be used to simulate traffic behavior in a city, evaluate the impact of different transportation routes on product delivery, or simulate the behavior of a distribution system.

Simulation in energy: Simulation is used in energy to model energy systems and predict their behavior in different situations. For example, it can be used to simulate the behavior of an electric grid, evaluate the impact of different energy sources on the environment, or simulate the behavior of an energy storage system.

Advantages and disadvantages of simulation in decision-making

Simulation has several advantages and disadvantages that should be taken into account when using it in decision-making.

Advantages:

Allows modeling of complex systems: Simulation allows modeling of complex systems that would be difficult to understand or predict using analytical methods.

Allows simulation of different scenarios: Simulation allows simulation of different scenarios and evaluating the impact of each one. This allows making informed decisions on how to approach different situations.

Reduces risk: Simulation allows reducing risk by predicting the behavior of a system before making a decision. This can help avoid costly or dangerous errors.

Saves time and money: Simulation can save time and money by allowing different scenarios to be tested before implementing a solution.

Disadvantages:

Requires precise data: Simulation requires precise and complete data to build an accurate model. If the data is inaccurate or incomplete, the model will not accurately reflect the behavior of the real system.

Requires technical knowledge: Simulation requires technical knowledge to build models and run simulations. If the necessary experience is not available, errors can be made that affect the accuracy of the model.

Does not always reflect reality: Simulation is a simplification of the real world and there is always a degree of uncertainty associated with simulation models. Additionally, the assumptions used in the model may not be accurate and can affect the accuracy of the result.

Can be costly: Building an accurate simulation model can be costly, especially if specialized software or high-end hardware is required.

Overall, simulation can be a powerful tool for decision-making, especially when faced with complex or uncertain situations. However, it is important to recognize that simulation has limitations and that the results obtained must be interpreted carefully.

Stages of the Simulation Process

The simulation process consists of several stages that must be followed to obtain accurate and meaningful results. Below are the five main stages of the simulation process.

Problem Definition: In the first stage of the simulation process, the problem to be solved is defined and the simulation objectives are established. It is important to define the scope of the simulation and the limits of the model, including the assumptions used and the variables to be considered.

Model Construction: In the second stage of the simulation process, the mathematical model that describes the behavior of the system to be simulated is constructed. This involves defining the variables and relationships that influence the behavior of the system and developing the necessary equations to represent them.

Model Validation: In the third stage of the simulation process, the constructed model is validated to ensure that it is accurate and accurately represents the behavior of the real system. This involves comparing the simulation results with actual data to determine the accuracy of the model.

Simulation Execution: In the fourth stage of the simulation process, the simulation is executed using the input data defined in the first stage. Multiple simulations can be performed to evaluate different scenarios and obtain significant statistical results.

Results Analysis: In the fifth and final stage of the simulation process, the results obtained from the simulation are analyzed to evaluate the performance of the system in different situations and make informed decisions. The results should be compared with the objectives established in the first stage and the accuracy and reliability of the model should be evaluated.

Simulation Tools

There are several simulation tools available that can be used to construct and execute simulation models. Below are some of the most common tools used in simulation.

Discrete Event Simulators: Discrete event simulators are used to simulate systems that change state at discrete moments in time. This means that events occur at specific times and the system remains in a constant state between events.

Process Simulators: Process simulators are used to simulate systems that continuously change state over time. This implies that events occur continuously and the system does not remain in a constant state between events.

Monte Carlo Simulators: Monte Carlo simulators are used to simulate complex and stochastic systems using statistical methods. This technique uses random numbers to simulate variability and uncertainty in the model and provides probabilistic results.

System Dynamics Simulators: System dynamics simulators are used to simulate complex and dynamic systems that have multiple feedbacks and causal relationships. This technique uses flow diagrams to represent the system and can model the complexity of the interaction between different variables.

Agent-based Simulators: Agent-based simulators are used to simulate systems that consist of multiple individual agents that interact with each other. This technique is commonly used in modeling social and economic systems, such as simulating financial markets and decision-making in groups.

Each simulation tool has its own strengths and weaknesses, and the choice of tool depends on the type of problem being solved and the available data.

Examples of Simulation Applications:

Simulation is used in a wide variety of fields, from engineering and science to economics and social sciences. Below are some examples of simulation applications in different fields.

Engineering: Simulation is commonly used in engineering to design and optimize complex systems, such as airplanes, cars, and power plants. Simulation is used to evaluate the performance of different designs and to identify areas for improving efficiency and safety.

Health Sciences: Simulation is used in health sciences to evaluate the effect of different treatments and therapies on patients. Simulation is also used to train healthcare professionals in emergency situations and to develop and test new medical devices.

Economics and Finance: Simulation is used in economics and finance to model the behavior of financial markets and evaluate different investment strategies. Simulation is also used to evaluate the effect of different economic policies and to predict the impact of future events on the economy.

Social Sciences: Simulation is used in social sciences to model the behavior of individuals and groups in different situations. Simulation is used to evaluate the effect of different public policies and to predict the impact of future events on society.

Conclusion

Simulation is a powerful decision-making tool that is used in a wide variety of fields. Simulation can be used to evaluate different scenarios and make informed decisions in complex and uncertain situations. However, it is important to recognize that simulation has limitations and that the results obtained should be interpreted carefully. The simulation process consists of several stages that must be followed to obtain accurate and meaningful results, and there are several simulation tools available that can be used to build and execute simulation models.

SUPPLY CHAIN AND LOGISTICS MANAGEMENT

Supply chain and logistics management is one of the most important aspects of success for any business. It refers to the planning, coordination, and control of activities related to the procurement, manufacturing, and delivery of products or services to customers. In this chapter, we will explore the key concepts of supply chain and logistics management, including demand planning, inventory management, inbound and outbound logistics, and supply chain optimization.

Demand Planning:

Demand planning is the process of predicting the quantity of products or services that customers will purchase in the future. The accuracy of this prediction is crucial to ensure that a company has enough products in stock to meet customer demand without incurring additional inventory costs. To conduct effective demand planning, companies can use a variety of techniques, including collecting historical sales data, conducting customer opinion surveys, and analyzing market trends.

Inventory Management:

Inventory management is the process of managing a company's inventory to ensure that there are enough products in stock to meet customer demand without incurring unnecessary storage and management costs. Effective inventory management requires a combination of demand planning, forecasting, and supply chain management techniques. Companies can use a variety of tools and technologies to optimize their inventory management, including inventory management software, data analysis, and real-time inventory tracking systems.

Inbound Logistics:

Inbound logistics refers to the process of receiving and managing the materials and components needed for the production of products or services. Effective inbound logistics is essential to ensure that products are produced and delivered on time and at a reasonable cost. Companies can optimize their inbound logistics by implementing production planning systems, supply chain tracking, and collaboration with suppliers to improve efficiency.

Outbound Logistics:

Outbound logistics refers to the process of managing and delivering products or services to end customers. Effective outbound logistics is essential to ensure that products are delivered on time and in good condition, which can have a significant impact on customer satisfaction and brand loyalty. Companies can optimize their outbound logistics by implementing real-time shipping tracking systems, collaborating with logistics partners, and using information technology to improve supply chain visibility and efficiency.

Supply Chain Optimization:

Supply chain optimization is the process of identifying and eliminating inefficiencies and redundancies in a company's supply chain. Supply chain optimization can significantly improve efficiency, reduce costs, and improve customer satisfaction. Companies can optimize their supply chain through the implementation of information technology, such as the use of supply chain management systems and data analysis. They can also collaborate with suppliers and logistics partners to improve efficiency and transparency in the supply chain.

In addition, supply chain optimization may involve reorganizing processes and eliminating bottlenecks, which can improve efficiency throughout the supply chain. For example, a company could use lean techniques to identify and eliminate waste in the production and delivery of products.

Another common strategy for supply chain optimization is outsourcing non-core activities. For example, a company may outsource inbound logistics or inventory management to a specialized third party, allowing the company to focus on its core competencies and reduce costs.

Supply chain optimization may also involve improving collaboration and communication throughout the supply chain. Companies can work with suppliers and logistics partners to improve visibility and transparency in the supply chain, which can help prevent problems and delays in product delivery.

Conclusion:

In summary, supply chain and logistics management are key aspects of business success. Demand planning, inventory management, inbound and outbound logistics, and supply chain optimization are fundamental aspects of supply chain and logistics management. Companies can use a variety of tools and techniques to improve their supply chain management, including the use of information technology, collaboration with suppliers and logistics partners, and implementation of supply chain optimization strategies. By focusing on supply chain and logistics management, companies can improve efficiency, reduce costs, and improve customer satisfaction.

DESIGN OF FACILITIES AND PLANT LAYOUT

The design of facilities and plant layout are fundamental to the success of any company. The way production areas are organized, the location of equipment, and the efficiency of processes have a direct impact on the quality of products, delivery times, and operating costs.

This chapter covers the main concepts and strategies related to the design of facilities and plant layout, with a focus on resource optimization and continuous process improvement.

Design of facilities:

Facility design is the process of planning and configuring the physical spaces where productive activities take place. This includes the location of production areas, equipment, raw materials, and finished products, as well as the definition of workflow and space organization.

One of the main objectives of facility design is to optimize resource utilization, which involves maximizing efficiency and productivity of production processes, minimizing production times, and reducing operating costs. To achieve this, it is necessary to consider a number of factors such as the size and shape of spaces, the location of equipment and storage areas, and the layout of electrical, hydraulic, and ventilation facilities, among others.

Efficiency in facility design can significantly improve production processes. For example, proper planning of work areas can reduce material and product transportation times, resulting in greater efficiency and productivity. Likewise, the arrangement of production equipment can facilitate tasks and minimize the risk of work accidents.

Another important aspect of facility design is adaptability. Facilities must be flexible and able to be adapted to the changing needs of the company, depending on changes in demand, the introduction of new products, or the adoption of new technologies.

Plant layout:

Plant layout is the process of organizing and arranging production areas within an industrial plant. The purpose of plant layout is to achieve efficient interaction between available resources and production processes.

The main objective of plant layout is to reduce costs and improve production efficiency. To achieve this, it is necessary to consider a number of factors, such as the location of equipment, the layout of storage areas, the organization of workflow, and space planning.

Plant layout can also impact product quality and customer satisfaction. Proper arrangement of work areas can reduce production times, allowing for faster delivery of products with higher quality.

There are several approaches to plant layout design, including the process approach, which focuses on grouping activities by function, and the product approach, which groups production activities around the products being manufactured. There is also a hybrid approach, which combines both approaches to obtain the benefits of each.

The process approach is used when different products are produced using the same equipment and processes. In this case, production areas are grouped based on the activity being performed, such as machining, assembly, or finishing. This approach maximizes process efficiency, as the same equipment can be used to produce different products.

The product approach, on the other hand, is used when different products require specific processes and equipment. In this case, production areas are organized based on the products being manufactured. This approach optimizes production for each product, as specific processes and equipment can be tailored to each product.

The hybrid approach combines both approaches to obtain the benefits of each. In this case, production areas are grouped based on processes, but are adapted to the specific needs of each product.

Factors to consider in facility design and plant layout:

When designing facilities and laying out plants, it is necessary to consider a number of factors that have a direct impact on process efficiency and company profitability. Some of the most important factors to consider are:

Production capacity: It is necessary to determine the production capacity of the facilities to ensure that they are sufficient to meet current and future demand.

Workflow: It is necessary to analyze workflows to ensure that processes are performed efficiently and without interruptions.

Safety: It is necessary to ensure the safety of workers and the integrity of equipment, through the implementation of appropriate safety measures.

Ergonomics: It is necessary to consider ergonomics in the design of facilities and plant layout to avoid injuries or fatigue in workers.

Energy efficiency: It is necessary to consider energy efficiency in the design of facilities and plant distribution to reduce operating costs.

Flexibility: It is necessary to ensure that facilities are flexible and can be adapted to the changing needs of the company.

Strategies for continuous improvement in facility design and plant distribution:

Continuous improvement in facility design and plant distribution is crucial to maintaining efficiency and profitability in the company. Some strategies for continuous improvement in facility design and plant distribution include:

Implementation of quality management systems: Implementing quality management systems, such as ISO 9001, can help the company identify improvement opportunities in facility design and plant distribution.

Periodic evaluation of facilities: It is necessary to conduct periodic evaluations of facilities to identify improvement opportunities in plant distribution, workflow, safety, and energy efficiency.

Technology upgrades: Upgrading technologies can help improve process efficiency and reduce operating costs in facility design and plant distribution.

Worker training and education: It is necessary to ensure that workers are trained to use facilities efficiently and safely through continuous training and education.

Implementation of continuous improvement processes: Implementing continuous improvement processes, such as Lean Manufacturing or Six Sigma, can help identify and eliminate waste in production processes and improve efficiency in facility design and plant distribution.

Collaboration with suppliers and customers: Collaboration with suppliers and customers can help identify improvement opportunities in facility design and plant distribution and improve efficiency throughout the supply chain.

Conclusions:

Facility design and plant distribution are crucial to maintaining efficiency and profitability in the company. When designing facilities and distributing plants, it is necessary to consider a range of factors that have a direct impact on process efficiency and company profitability, such as production capacity, workflow, safety, ergonomics, energy efficiency, and flexibility.

Continuous improvement in facility design and plant distribution is crucial to maintaining efficiency and profitability in the company. Some strategies for continuous improvement in facility design and plant distribution include implementation of quality management systems, periodic evaluation of facilities, technology upgrades, worker training and education, implementation of continuous improvement processes, and collaboration with suppliers and customers.

In summary, facility design and plant distribution are critical tasks for the efficiency and profitability of the company. By considering the factors mentioned and implementing continuous improvement strategies, companies can significantly improve process efficiency and reduce operating costs, resulting in a competitive advantage in the market.

PRODUCTION PLANNING AND SCHEDULING

The chapter on Production Planning and Scheduling is one of the most important topics in production management, as it is responsible for establishing the necessary processes to achieve efficiency and effectiveness in the production of goods and services. This chapter covers the most relevant aspects of production planning and scheduling, its importance in business management, the methods and tools used, as well as the factors that influence its development and application.

Importance of Production Planning and Scheduling

Production planning and scheduling is essential for efficient production management. Its objective is to ensure that resources are available when needed to produce the required goods and services in the appropriate quantity and quality, at the lowest possible cost. This improves the profitability and competitiveness of the company.

Production planning and scheduling is a continuous process that begins with the definition of the company's objectives and goals, and extends to detailed production scheduling in the short term. Long-term objectives include defining the production strategy, identifying the markets and customers targeted by the company, defining the products and services offered, identifying the necessary resources, and planning production capacity.

In the short term, production planning and scheduling involves detailed scheduling of daily or weekly production, resource allocation to production tasks, production monitoring and control, and problem resolution and deviation management.

Methods and Tools for Production Planning and Scheduling

There are various methods and tools for production planning and scheduling, which are selected based on the characteristics and needs of the company and production. Some of the most commonly used methods and tools are:

Capacity Planning: This method involves evaluating available production capacity and identifying the resources needed to meet established production objectives. This allows the company to plan investment in resources necessary to increase production capacity and improve efficiency.

Production Scheduling: This tool allows for the allocation of resources to production tasks and the definition of start and end dates for activities. Production scheduling is carried out through computer tools such as production management systems (ERP) and advanced scheduling tools.

Production Control: Production control involves monitoring the progress of production and identifying deviations and problems. To do this, production tracking and control tools are used, such as Key Performance Indicators (KPIs), quality management systems, and Statistical Process Control (SPC) systems.

Factors Affecting Production Planning and Scheduling

Production planning and scheduling can be affected by various factors that influence its development and application. Some of these factors include:

Market demand: Market demand is a critical factor that affects production planning and scheduling. If market demand is high, the company must plan and schedule its production efficiently to ensure that it can meet market needs in the shortest possible time. On the other hand, if demand is low, the company must plan and schedule its production in a way that does not generate excess inventory that can result in unnecessary costs.

Resource availability: Resource availability, such as materials, machinery, and personnel, can influence production planning and scheduling. If resources are not available when needed, it can cause production delays and affect the company's ability to meet its delivery commitments.

Technical limitations: Technical limitations can influence production planning and scheduling. For example, if machinery cannot produce a certain amount of goods in a given time, the company must adjust its production plan accordingly.

Seasonality: Seasonality is an important factor that influences production planning and scheduling. For example, a company that produces toys may have very high demand during the holiday season and very low demand for the rest of the year. The company must plan and schedule its production to be able to meet demand during periods of high demand and avoid excess inventory during periods of low demand.

Competition: Competition is an important factor that influences production planning and scheduling. If competitors offer similar products at lower prices, the company must plan and schedule its production in a way that can compete on price and quality.

In summary, production planning and scheduling is a fundamental process in business management, as it allows for the optimization of resources and the improvement of efficiency and effectiveness in the production of goods and services. To achieve effective planning and scheduling, various factors that can affect the process must be considered, and appropriate tools and methods must be used to enable efficient and effective production management.

INVENTORY AND WAREHOUSE MANAGEMENT

Inventory and warehouse management is one of the most important activities in any company that sells products. Inventories represent a significant investment for the company and proper management can maximize profits and minimize losses. Warehouse management is the process of managing and organizing the products stored in a particular location. In this chapter, we will discuss the main strategies for inventory and warehouse management.

Inventory Management Strategies

Inventory management is the process of controlling the flow of products in a company. It is important to ensure that products are available to customers at the right time and in the right quantity. Below are the main strategies for inventory management.

Just-In-Time (JIT)

The JIT strategy involves having inventory just when it is needed. This means that the company does not maintain large quantities of inventory in stock, which reduces storage costs. The JIT strategy can be very effective for companies that produce high-demand products, as it minimizes inventory cost without sacrificing production capacity. However, this strategy is highly dependent on the supplier's ability to deliver products on time, which can be a risk.

Maximum-Minimum

The maximum-minimum strategy is based on establishing a maximum and minimum amount of inventory that must be maintained. When inventory reaches the minimum point, a replenishment order is placed. This strategy is useful for products that have a stable demand, as it allows for minimum inventory without risking running out of products. However, this strategy can lead to excess inventory if appropriate levels are not established.

ABC

The ABC strategy is based on classifying products into three categories: A, B, and C. A products are those with high demand and represent a large part of the company's sales. B products have moderate demand and C products have low demand. The company can apply different inventory management strategies for each category. For example, A products may have a higher inventory level than B and C products. This strategy allows the company to focus on the most important products and avoid excess inventory of low-demand products.

Warehouse Management Strategies

Warehouse management is the process of organizing the products stored in a particular location. Good warehouse management can improve efficiency and reduce storage costs. Below are the main strategies for warehouse management.

Fixed Location Storage

The fixed location storage strategy involves assigning a specific place for each product in the warehouse. This allows for more efficient inventory management and reduces the time needed to find a product. Additionally, this strategy can be very useful for companies that have a large number of different products.

Random Location Storage

The strategy of random location storage involves not assigning a specific location for each product in the warehouse. Instead, products are placed in any available space at the time they arrive at the warehouse. This strategy can be very effective for companies with a high product turnover rate, as it reduces the time required to find a specific product. However, this strategy can be more challenging to manage if the company has a large number of different products.

Cross-docking

The cross-docking strategy involves receiving products and directly shipping them to customers without storing them in the warehouse. This strategy can be very effective for companies with a high product turnover rate and an efficient distribution system. However, this strategy requires careful planning and efficient coordination between suppliers and customers.

Temperature Storage

The temperature storage strategy involves storing products at different temperatures according to their specific requirements. For example, perishable products may require cold storage, while electronic products may require dry and temperature-controlled storage. This strategy can be very effective for companies that sell a wide variety of products with different storage requirements.

Automated Storage

The automated storage strategy involves the use of automated systems to store and retrieve products in the warehouse. These systems can include robots, conveyors, and automated inventory control systems. This strategy can improve the efficiency and accuracy of the warehouse management process, but requires a significant investment in technology and training.

Conclusion

Inventory and warehouse management is a critical activity for any company engaged in the sale of products. Proper inventory management can maximize profits and minimize losses, while proper warehouse management can improve efficiency and reduce storage costs. The inventory and warehouse management strategies described in this chapter can help companies optimize their inventory and warehouse management according to their specific needs and requirements. It is important for companies to carefully evaluate these strategies and choose the ones that best fit their operations and business objectives.

METHODS AND TIME ENGINEERING

Methods and time engineering is a discipline that focuses on improving production processes through the identification and elimination of unnecessary activities and the optimization of necessary tasks to complete an activity or project. In this chapter, we will explain in detail what methods and time engineering involves, its objectives, its main techniques and tools, as well as its importance in the management of production processes and continuous improvement of organizations.

Definition of methods and time engineering

Methods and time engineering is defined as a discipline whose main objective is to improve production processes through the identification and elimination of unnecessary activities and the optimization of necessary tasks to complete an activity or project. This discipline focuses on efficiency management, cost reduction, and waste elimination in production processes.

Methods and time engineering is based on the systematic study of production processes to identify and eliminate all activities that do not add value, and optimize activities that do. Specific techniques and tools are used to measure, analyze and improve production processes.

Objectives of methods and time engineering

The objectives of methods and time engineering are as follows:

Efficiency improvement: Methods and time engineering aims to improve the efficiency of production processes by reducing the time required to complete an activity or project.

Cost reduction: Improving efficiency reduces the costs associated with completing an activity or project by eliminating unnecessary activities and optimizing necessary tasks.

Waste elimination: Methods and time engineering aims to eliminate waste in production processes, resulting in cost reduction and improvement in the quality of the product or service.

Quality improvement: Improving efficiency and eliminating waste leads to quality improvement in the product or service by eliminating errors and optimizing necessary tasks.

Techniques and tools of methods and time engineering

The main techniques and tools used in methods and time engineering are as follows:

Flowcharts: Flowcharts represent production processes graphically, identifying necessary activities and their relationship. These charts are useful for identifying unnecessary activities and optimizing necessary tasks.

Process analysis: Process analysis identifies and eliminates all activities that do not add value and optimizes necessary tasks. This analysis is conducted through direct observation of the production process and collection of relevant data, such as the time taken to complete each activity.

Time and motion study: The time and motion study is a technique that allows measuring the time required to perform a task or activity, identifying the necessary movements and eliminating those that do not add value. This technique is useful for improving efficiency and reducing costs associated with carrying out an activity or project.

Line balancing: Line balancing is a technique that allows optimally distributing the tasks necessary for the completion of a project or activity, in order to maximize efficiency and reduce costs. This technique is especially useful in chain production processes, where each task depends on the completion of the previous one.

Continuous improvement: Continuous improvement is a philosophy that focuses on the constant search for the improvement of production processes, through the identification and elimination of unnecessary activities and the optimization of necessary tasks. This philosophy is based on the belief that it is always possible to improve, and that continuous improvement is essential for the survival and growth of organizations.

Importance of methods and time engineering

Methods and time engineering is essential for the management of production processes and the continuous improvement of organizations, as it allows identifying and eliminating all activities that do not add value, and optimizing necessary tasks. Some of the reasons why methods and time engineering is important are:

Cost reduction: Improving efficiency and eliminating waste allows reducing costs associated with carrying out an activity or project, which translates into increased profitability for the organization.

Improvement in quality: Improving efficiency and eliminating errors allows improving the quality of the product or service, which translates into greater customer satisfaction and improvement in the reputation of the organization.

Greater flexibility: Optimizing production processes allows for greater flexibility in production management, which allows adapting to market needs and changes in demand.

Greater competitiveness: Improving efficiency and reducing costs allows for greater competitiveness in the market, which translates into a greater market share and an increase in the organization's profits.

Conclusions

In conclusion, methods and time engineering is an essential discipline for the management of production processes and the continuous improvement of organizations. This discipline allows identifying and eliminating all activities that do not add value, and optimizing necessary tasks for carrying out an activity or project. Techniques and tools used in methods and time engineering, such as flow charts, process analysis, time and motion study, line balancing, and continuous improvement, are essential for improving efficiency, reducing costs, improving quality, and competitiveness of organizations.

It is important to note that methods and time engineering is not a technique used only in the manufacturing industry but is also applicable in other areas, such as project management, logistics, customer service, among others. In any activity where a task needs to be performed, methods and time engineering can be applied to improve its efficiency and reduce associated costs.

In short, methods and time engineering is a discipline that provides great benefits to organizations, allowing for the improvement of production processes, cost reduction, and quality improvement, which translates into greater customer satisfaction and competitiveness in the market. Therefore, it is essential that organizations invest in the training and development of their teams in this discipline, to take advantage of its full potential and continue growing and improving in the future.

ERGONOMICS AND OCCUPATIONAL SAFETY

Ergonomics and occupational safety are two fundamental areas in the workplace. Ergonomics studies the relationship between the worker and their work environment with the aim of improving efficiency, productivity, and the worker's quality of life. On the other hand, occupational safety focuses on identifying, evaluating, and controlling risks that can affect the health and safety of workers.

This chapter will address the main aspects related to ergonomics and occupational safety. Basic concepts of ergonomics will be reviewed, as well as the main techniques and tools used for its application in the workplace. Additionally, the main regulations related to occupational safety will be analyzed, and the main strategies for preventing and controlling occupational risks will be discussed.

Ergonomics in the workplace

Ergonomics is a discipline that studies the relationship between the worker and their work environment with the aim of optimizing the worker's well-being, increasing efficiency and productivity, and reducing the risk of occupational injuries and illnesses.

The first step to applying ergonomics in the workplace is to conduct a detailed analysis of the tasks and activities carried out in the company. This includes observing workers in their work environment, as well as evaluating the equipment, tools, and furniture used.

Once the analysis has been conducted, the following ergonomics techniques and tools can be applied in the workplace:

Ergonomic design of the workstation: The ergonomic design of the workstation involves adapting the work environment to the worker with the aim of minimizing the risk of occupational injuries and illnesses. This includes selecting ergonomic furniture and equipment, optimizing lighting and temperature, and adapting the workspace to meet the needs of the worker.

Workload assessment: Workload assessment involves identifying tasks and activities that require greater physical or mental effort with the aim of designing a work environment that reduces fatigue and stress in the worker.

Biomechanical analysis: Biomechanical analysis involves evaluating how the human body moves and exerts itself in different work situations with the aim of designing a work environment that minimizes the risk of musculoskeletal injuries.

Posture analysis: Posture analysis involves evaluating how the worker sits, stands, or moves in the work environment with the aim of designing a work environment that promotes a healthy posture and reduces the risk of musculoskeletal injuries.

Ergonomics training: Ergonomics training involves educating workers about the basic principles of ergonomics with the aim of promoting a healthy work culture and preventing occupational injuries. Workers can learn about the importance of proper posture, how to adjust the height of the desk and chair, how to use ergonomic equipment, and how to take breaks to stretch and rest.

Safety regulations and standards in the workplace

Workplace safety is a discipline that focuses on identifying, evaluating, and controlling risks that may affect the health and safety of workers. To this end, there are various regulations and standards that establish the obligations and responsibilities of employers and workers in the field of workplace safety.

Some of the most important workplace safety regulations and standards are:

Regulations on Health and Safety at Work: These regulations establish the obligations and responsibilities of employers and workers in the field of workplace safety. It also sets criteria for the identification, evaluation, and control of occupational risks.

Regulations on Safety and Health at Work: These regulations establish provisions for the prevention of occupational accidents and illnesses. They also set out procedures and criteria for evaluating occupational risks and implementing preventive measures.

Strategies for preventing and controlling occupational risks

Preventing and controlling occupational risks is essential to ensure a healthy and safe work environment. There are various strategies that can be implemented by employers and workers, including:

Risk assessment and management: Risk assessment and management involves identifying occupational risks, evaluating their likelihood of occurrence and severity, and implementing preventive measures to minimize or eliminate these risks.

Training and education: Training and education of workers on workplace safety are essential to promote a culture of safe work. Workers should be trained in the use of personal protective equipment, identifying occupational risks, and implementing preventive measures.

Implementation of safety measures: Safety measures are technical or administrative measures that are implemented to reduce or eliminate occupational risks. These measures may include the installation of collective protection equipment, the implementation of safe work procedures, and the use of personal protective equipment.

Safety inspections: Safety inspections are an important tool for identifying occupational risks and evaluating the effectiveness of preventive measures implemented. Inspections can be carried out by employers or workers and should be recorded and reported for follow-up.

Conclusions

Ergonomics and occupational safety are fundamental aspects for ensuring a healthy and safe work environment. Ergonomics focuses on the design and adaptation of the work environment to the physical and psychological needs of workers, while occupational safety focuses on identifying, evaluating, and controlling risks that may affect the health and safety of workers.

It is important for employers and workers to be aware of the importance of ergonomics and occupational safety and work together to implement preventive measures and ensure a safe work environment. This involves identifying and evaluating occupational risks, implementing preventive measures, providing training and education to workers, and conducting safety inspections.

Additionally, it is important for employers to comply with the safety and health regulations established by competent authorities. The General Law on Health and Safety at Work, the NOMs, and the Federal Regulation on Safety and Health at Work establish the obligations and responsibilities of employers and workers in terms of occupational safety and establish criteria for identifying, evaluating, and controlling occupational risks.

In conclusion, ergonomics and occupational safety are fundamental aspects for ensuring a healthy and safe work environment. Employers and workers must work together to implement preventive measures and ensure a safe work environment and comply with the safety and health regulations established by competent authorities.

MAINTENANCE MANAGEMENT AND RELIABILITY

Maintenance management and reliability are critical aspects for the success of any organization. The effective implementation of these processes can improve operational efficiency, reduce maintenance costs, increase asset lifespan, and enhance personnel and machinery safety. In this chapter, we will explore the basics of maintenance management and reliability, their benefits, and how they can be implemented in an organization.

Maintenance Management

Maintenance management refers to the processes used to maintain and improve the functionality, reliability, and safety of equipment and systems in an organization. Maintenance management can be divided into two main categories: corrective maintenance and preventive maintenance.

Corrective maintenance is performed after a piece of equipment or system has failed. The main objective of corrective maintenance is to repair the equipment or system as quickly as possible to minimize downtime. However, corrective maintenance can be expensive and may have a negative impact on an organization's productivity and profitability.

Preventive maintenance, on the other hand, is performed before a failure occurs in equipment or system. The main objective of preventive maintenance is to prevent failures in equipment or system and minimize downtime. Preventive maintenance can be divided into two categories: time-based maintenance and condition-based maintenance.

Time-based maintenance is performed at regular intervals, regardless of the equipment or system's condition. This type of maintenance is useful for equipment and systems that have a predictable lifecycle, such as changing a car's oil after a certain number of kilometers.

Condition-based maintenance, on the other hand, is performed when the equipment or system reaches a certain level of degradation or when an anomaly is detected. This type of maintenance is based on continuous monitoring of equipment and systems to identify issues before failures occur.

Reliability

Reliability refers to the ability of equipment or system to operate continuously and without failures for a certain period of time. Reliability is a critical aspect of any equipment or system, especially in high-security and mission-critical environments. Reliability can be improved by implementing maintenance management processes and applying reliability analysis techniques.

Failure rate is a common measure of reliability. Failure rate refers to the frequency with which failures occur in equipment or system during a certain period of time. Failure rate can be reduced by implementing preventive maintenance processes, using high-quality materials, and improving equipment or system design.

The Mean Time Between Failures (MTBF) is another common measure of reliability in industrial engineering. The MTBF refers to the average time between two consecutive failures in a piece of equipment or system. The MTBF can be increased by improving the equipment or system design, using high-quality materials, and implementing preventive maintenance processes.

Another important aspect of reliability is availability. Availability refers to the time during which a piece of equipment or system is available and operational. Availability can be improved by implementing preventive maintenance processes, reducing repair time, and improving the equipment or system design.

Implementation of maintenance and reliability management

The effective implementation of maintenance and reliability management can significantly improve operational efficiency, reduce maintenance costs, and enhance personnel and machinery safety. Here are some key steps for successfully implementing maintenance and reliability management in an organization:

Evaluate the current situation: The first step in implementing maintenance and reliability management is to evaluate the current situation of the organization. This may include reviewing existing maintenance records, identifying current problems, and assessing equipment or system performance.

Develop a maintenance plan: Once the current situation has been evaluated, it is important to develop a maintenance plan that addresses identified problem areas. The maintenance plan should include a combination of preventive and corrective maintenance to minimize downtime and improve equipment or system reliability.

Implement preventive maintenance processes: Preventive maintenance is fundamental to improving equipment or system reliability. Preventive maintenance processes should be implemented to ensure that equipment or systems are kept in optimal condition.

Use reliability analysis techniques: Reliability analysis techniques such as Failure Mode and Effect Analysis (FMEA) and Fault Tree Analysis (FTA) can help identify areas of highest risk and develop mitigation strategies.

Train personnel: Personnel should be trained in maintenance processes and reliability analysis techniques. This will ensure that personnel are prepared to perform preventive maintenance and identify potential problems before failures occur.

Monitor performance: Once maintenance and reliability management processes are implemented, it is important to monitor equipment or system performance. This may include monitoring failure rates, MTBF, and availability.

Benefits of Maintenance and Reliability Management

The effective implementation of maintenance and reliability management can provide several significant benefits for an organization. Some of these benefits include:

Reduction of maintenance costs: The implementation of preventive maintenance processes can reduce long-term maintenance costs by minimizing the need for costly and extensive repairs.

Improvement of operational efficiency: The implementation of preventive maintenance processes can improve operational efficiency by minimizing downtime and increasing equipment or system availability.

Increase in asset lifespan: The implementation of preventive maintenance processes can increase asset lifespan by minimizing equipment or system wear and fatigue.

Improvement of safety: The implementation of preventive maintenance processes can improve personnel and machinery safety by identifying and correcting potential issues before failures occur.

Improvement of product quality: The implementation of preventive maintenance processes can improve product quality by reducing variability in the production process and minimizing product defects.

Improvement of customer satisfaction: Improving operational efficiency and product quality can improve customer satisfaction by providing high-quality products in a faster delivery time.

Conclusions

In summary, maintenance and reliability management are fundamental to operational efficiency, safety, and product quality. The effective implementation of maintenance and reliability management can reduce maintenance costs, improve operational efficiency, increase asset lifespan, improve safety, improve product quality, and improve customer satisfaction. Preventive maintenance processes, reliability analysis techniques, and personnel training are some of the key aspects of maintenance and reliability management. By implementing these processes, organizations can achieve greater reliability and operational efficiency in their production.

AUTOMATION AND CONTROL TECHNOLOGIES

Currently, automation and control technology is present in a wide variety of industries, from manufacturing production to the management of infrastructure and public services. This technology uses control systems to automate processes and improve the efficiency, quality, and safety of the products and services offered.

In this chapter, we will explore the main automation and control technologies, as well as the benefits and challenges they present. In addition, we will examine the different types of control systems, from relay-based control systems to more advanced programmable control systems.

Automation and Control Technologies

Automation and control technologies include a wide range of tools and systems designed to improve the efficiency, quality, and safety of industrial processes. These technologies are used in a variety of industries, from manufacturing to the management of infrastructure and public services.

Some of the main automation and control technologies include:

Computer Numerical Control (CNC)

Computer Numerical Control (CNC) is an automation system used in manufacturing production to control machine tools using software programs. CNC software programs allow users to define the tools and movements that should be performed on the workpiece.

CNC is widely used in the production of high-precision parts, such as components for the aerospace and medical industries.

Robotics

Robotics is an automation technology that uses robots to perform tasks in a variety of industrial environments. Robots can be programmed to perform repetitive and dangerous tasks, which increases worker safety and improves process efficiency.

Robotics is used in a wide range of industries, from manufacturing to logistics and healthcare.

Process Control Systems

Process control systems are automation tools used to control and monitor industrial processes in real time. These systems use sensors and controllers to measure and adjust process parameters, such as temperature, pressure, and flow.

Process control systems are used in a wide range of industries, from chemical production to food and beverage manufacturing.

Quality Control Systems

Quality control systems are automation tools used to ensure that products and services meet required quality standards. These systems use measurement and analysis techniques to detect and correct quality problems.

Quality control systems are used in a wide range of industries, from manufacturing to healthcare and financial services.

Types of Control Systems

There are several types of control systems used in automation and control technology. These systems are classified based on their complexity and programming capability.

Relay-based Control Systems

Relay-based control systems are the oldest and simplest control systems. These systems use electromechanical relays to control the operation of a machine or process. Relays are activated or deactivated based on the electrical signal they receive, allowing for control of the on/off operation of system components.

Relay-based control systems are limited in their programming capabilities and can only perform simple and repetitive tasks. However, these systems are reliable and are used in applications where simplicity is more important than advanced functionality.

Programmable Logic Controller (PLC)

Systems Programmable logic controller (PLC) systems are a more advanced form of control that uses a programmable computer to control the operation of a machine or process. PLCs use specialized programming language to control the sequence of operations and adjust process parameters.

PLCs are flexible and can be programmed to perform a wide variety of tasks and operations. These systems are used in a wide range of industrial applications, from manufacturing production to management of infrastructure and public services.

Distributed Control Systems (DCS)

Distributed control systems (DCS) are advanced control systems used in complex industrial processes. These systems use a network of distributed controllers to monitor and control multiple processes in real-time.

DCSs are highly scalable and can be programmed to control large-scale and complex processes. These systems are used in a wide range of industrial applications, from chemical production to power generation and infrastructure management.

High-Level Control Systems (HIL)

High-level control systems (HIL) are simulation systems used to test and validate control systems in virtual environments. These systems use mathematical models to simulate system behavior in real-time, allowing engineers to test and optimize control system performance before implementing them in a real-world environment.

HIL systems are used in a wide range of industrial applications, from automotive to aerospace and electrical engineering.

Benefits of Automation and Control Technology

Automation and control technology offer several benefits to companies and organizations that use it. Some of the main benefits include:

Increased Efficiency

Automation and control allow companies to improve the efficiency of their processes, reducing production time and minimizing human errors. This allows companies to increase their production and reduce manufacturing costs.

Improved Quality

Automation and control technology allows companies to improve the quality of their products and services, reducing variability and minimizing errors. This increases customer satisfaction and improves the brand image.

Increased Safety

Automation and control technology allows for improved workplace safety by minimizing human exposure to hazardous situations. This is achieved by automating hazardous processes or by implementing safety systems that monitor the operation of machines and equipment to detect faults or dangerous conditions.

Cost Reduction

Automation and control technology can help reduce production costs by minimizing waste of materials and energy, as well as reducing the need for human labor.

Increased Flexibility

Advanced control systems, such as PLCs and DCSs, are highly flexible and can be programmed to perform a wide variety of tasks and operations. This allows companies to quickly adapt to changes in the market and new customer demands.

Improved Data Analysis

Automation and control technology allows companies to collect and analyze large amounts of data about their processes and operations. This enables them to identify areas for improvement and optimize their processes to maximize efficiency and profitability.

Challenges in Implementing Automation and Control Technology

While automation and control technology offers a number of benefits to companies, it also presents a number of challenges in its implementation. Some of the main challenges include:

Cost

The implementation of advanced control systems can be costly, especially for small and medium-sized enterprises. In addition to the cost of equipment and control systems, highly trained personnel are also required to install, program, and maintain these systems.

Integration

The implementation of advanced control systems often requires the integration of multiple systems and equipment, which can be a technical challenge. In addition, control systems may not be compatible with existing systems, requiring additional investment in updating existing systems or acquiring new equipment.

Personnel Training

The implementation of advanced control systems requires highly trained and specialized personnel to program, maintain, and operate the systems. Personnel training can be costly and time-consuming, which can delay the implementation of the control system.

Cybersecurity

Advanced control systems are connected to the network and can be vulnerable to cyber attacks. Cybersecurity must be an important consideration in the implementation of advanced control systems to protect systems and sensitive data.

Examples of Automation and Control Technology in Industry

Automation and control technology is used in a wide range of industries, from manufacturing to energy and automotive. The following are some examples of how automation and control technology is used in industry.

Automotive

In the automotive industry, automation and control technology is used to improve efficiency and quality in car production. Advanced control systems are used to control the production process, from part assembly to painting and final finishing. Robot and automated machine systems are also used to perform tasks that previously required human labor, such as component assembly and welding.

Energy

In the energy industry, automation and control technology is used to improve efficiency and safety in energy production, from electricity generation to oil and gas extraction. Advanced control systems are used to monitor and control the operation of power plants, which helps optimize production and reduce costs.

Manufacturing

In the manufacturing industry, automation and control technology is used to improve efficiency and quality in the production of goods, from electronics manufacturing to food and beverage production. Advanced control systems are used to monitor and control production processes, which helps to reduce costs and improve product quality.

Mining

In the mining industry, automation and control technology is used to improve safety and efficiency in the extraction of minerals and metals. Advanced control systems are used to monitor and control the operation of mining equipment, which helps to minimize risks for workers and maximize production.

Agriculture

In the agricultural industry, automation and control technology is used to improve efficiency and productivity in the production of food and crops. Advanced control systems are used to monitor and control processes such as irrigation, fertilization, and harvesting, which helps to maximize crop yields and reduce production costs.

Conclusions

Automation and control technology is a powerful tool for improving efficiency, safety, and profitability in a wide range of industries. The implementation of advanced control systems can help companies to reduce costs, improve product quality, and quickly adapt to changes in the market.

However, the implementation of advanced control systems also presents a number of challenges, such as cost, integration, personnel training, and cyber security. It is important for companies to carefully evaluate the benefits and challenges of implementing advanced control systems before making a decision.

Ultimately, automation and control technology can be a valuable tool to help companies remain competitive in an increasingly demanding global market.

INDUSTRIAL PROJECT DEVELOPMENT AND MANAGEMENT

The development and management of industrial projects is a crucial task for companies seeking to improve their production processes and, therefore, increase their competitiveness in the market. This chapter aims to provide an overview of the processes and tools necessary to successfully carry out industrial projects.

Phases of industrial project development

The development of an industrial project consists of several phases, which may vary depending on the nature of the project and the company carrying it out. However, it is possible to identify some phases that are common to most projects:

Identification of the need: In this phase, the problem or need to be solved through the project is defined. The project's objectives are also identified, and an estimated timeframe and budget are established.

Planning: In this phase, a detailed plan is developed that establishes the activities necessary to achieve the project's objectives, the resources that will be needed, and the deadlines in which they will be carried out.

Execution: In this phase, the planned activities are carried out, and the results obtained are monitored and controlled.

Closure: In this phase, the results obtained are evaluated, the project is documented, and the necessary actions are taken to close it.

Tools for industrial project management

There are various tools that can be used to carry out the management of industrial projects. Some of the most common ones are:

Gantt chart: This tool allows the graphical representation of the project's activity plan, showing the tasks to be performed and their estimated duration. It also allows the establishment of dependency relationships between tasks.

Pert Network: This tool is similar to the Gantt chart, but focuses on the dependency relationships between tasks. It allows identifying which are the critical tasks of the project, i.e., those that must be completed on time for the project to be completed on time.

Risk matrix: This tool allows identifying the risks associated with the project and establishing strategies to minimize or eliminate them. It also allows evaluating the probability and impact of each risk.

Contingency plan: This plan establishes the measures that will be taken in case of any unforeseen event that affects the project's development. It is important that this plan is established before the unforeseen event occurs, to be able to act quickly and effectively.

Project management software: There are various computer tools that allow the efficient management of projects, such as Microsoft Project or Trello. These tools allow planning tasks, establishing deadlines and resources, and monitoring the project's development in real-time.

Key factors for the success of an industrial project

The success of an industrial project depends on various factors, some of which are:

Clear definition of the project's objectives: It is important that the project's objectives are well defined from the beginning, so that all team members work in the same direction and know what the expected outcome is.

Proper resource allocation: it is crucial to assign the appropriate resources to the project, including personnel, budget, and time. Failure to allocate necessary resources may cause difficulties and delays in the project.

Effective communication: it is essential to have clear and effective communication among all members of the project team, as well as with clients and suppliers. Good communication can prevent misunderstandings and problems that could affect the project's success.

Risk management: it is important to identify the risks associated with the project and establish strategies to minimize or eliminate them. This can help prevent problems and delays in the project's development.

Monitoring and evaluation: it is necessary to constantly monitor the project to ensure that it is progressing according to the established plan and to detect any deviations or issues. Additionally, it is important to conduct an evaluation at the end of the project to determine if objectives have been met and expected results have been obtained.

Example of Industrial Project Management

To illustrate the management of an industrial project, the following hypothetical example is presented:

Company X wants to implement a new production line to manufacture a specific product. The project will be completed within six months, and a budget of $100,000 has been established. The project team consists of a project manager, a production engineer, an industrial designer, and a team of production operators.

Phase 1: Needs identification

The project team meets to define the problem they want to solve and the project's objectives. It is established that the company needs a new production line to manufacture a specific product due to increased demand. The project's objectives are to increase production of the product, improve quality, and reduce costs.

Phase 2: Planning

The project team develops a detailed plan that establishes the activities necessary to achieve the project's objectives, the resources required, and the deadlines for completion. The following main activities are established:

Production line design

Acquisition of necessary equipment

Production operator training

Production testing and adjustments

The project team uses a Gantt chart to graphically represent the project's activity plan and establish task dependencies.

Phase 3: Execution

The project team begins to carry out the planned activities and tracks and controls the results obtained. Regular meetings are established to assess project progress and make adjustments if necessary.

Phase 4: Closure

Once the implementation of the new production line is completed, the project is evaluated to determine if objectives have been met and expected results have been obtained. The project is documented, and closure is completed.

Conclusion

Managing industrial projects is a complex task that requires detailed planning, proper resource management, effective communication, risk identification and management, and constant project monitoring. The success of the project will depend largely on proper management of these key areas.

It is important for industrial project managers to have a good understanding of the activities and processes that will be carried out in the project, as well as the resources required to complete it. They also need leadership, communication, and time management skills to lead the project team and ensure that deadlines and objectives are met.

The use of project management tools, such as Gantt charts, can be very helpful in planning and monitoring industrial projects. These tools provide a clear visualization of activities and timelines and can help identify bottlenecks and delays in the project.

In summary, managing industrial projects is a complex process that requires careful planning, proper resource and risk management, effective communication, and constant project monitoring. Proper implementation of these practices can help ensure project success and client satisfaction.

COST-BENEFIT ANALYSIS AND PROJECT EVALUATION

In project management, it is essential to conduct a careful evaluation of costs and benefits before making any significant decisions. This is because any investment in a project must be profitable and beneficial to the organization as a whole. To evaluate the costs and benefits of a project, a technique called cost-benefit analysis is used. In this chapter, we will discuss in detail what cost-benefit analysis is, how it is carried out, and its importance in project evaluation.

What is Cost-Benefit Analysis?

Cost-benefit analysis is a technique used to evaluate the relationship between the costs and benefits of a project. This analysis is carried out to determine whether the expected benefits of the project justify the costs associated with it. In other words, cost-benefit analysis helps organizations determine whether a project is profitable and economically viable.

Cost-benefit analysis involves the identification of all costs associated with the project, including both direct and indirect costs. Direct costs are those that are directly related to the execution of the project, such as material and labor costs. Indirect costs, on the other hand, are those that are related to the project but are not directly attributable to it, such as general administrative costs.

On the other hand, project benefits must be identified and quantified. Benefits can be tangible or intangible. Tangible benefits are those that can be measured in monetary terms, such as increased sales or cost reduction. Intangible benefits are those that cannot be easily measured in monetary terms, such as brand image improvement or customer satisfaction.

Once all costs and benefits associated with the project have been identified, an analysis is performed to determine whether the benefits justify the costs. If the analysis shows that the benefits outweigh the costs, then the project is considered economically viable.

Steps to Conduct Cost-Benefit Analysis

Cost-benefit analysis is carried out in several stages. These stages include:

Identification of Costs and Benefits: The first step in cost-benefit analysis is to identify all costs and benefits associated with the project. This includes both direct and indirect costs and tangible and intangible benefits.

Quantification of Costs and Benefits: The next step is to quantify all identified costs and benefits. Costs are quantified in monetary terms, while tangible benefits are also quantified in monetary terms. Intangible benefits are quantified using subjective methods such as surveys or opinion analysis.

Establishment of a Baseline: Once all costs and benefits have been identified and quantified, a baseline is established. The baseline is a representation of the expected costs and benefits without the project in question. This helps to compare the expected costs and benefits with the actual costs and benefits after project implementation.

Cost-Benefit Analysis: The next step is to carry out a real cost-benefit analysis. This involves comparing the expected costs and benefits with the established baseline. If the expected benefits outweigh the expected costs, then the project is considered economically viable. If the expected costs outweigh the expected benefits, then the project may need to be re-evaluated or abandoned.

Importance of Cost-Benefit Analysis in Project Evaluation

Cost-benefit analysis is an important tool for project evaluation for several reasons:

Helps make informed decisions: Cost-benefit analysis provides important information about the relationship between the costs and benefits of a project.

This helps decision-makers to make informed decisions about whether to proceed with a project or not.

Identifies hidden costs: Cost-benefit analysis helps to identify hidden costs associated with a project. This includes indirect costs that are often overlooked but can have a significant impact on the profitability of the project.

Improves project planning: Cost-benefit analysis helps organizations to plan and budget a project effectively. This helps to avoid unpleasant surprises and ensure that project objectives are achieved within the allocated budget.

Reduces risk: Cost-benefit analysis helps organizations to evaluate the risk associated with a project. This includes the assessment of financial, operational, and compliance risks. If a high risk is identified, then steps can be taken to reduce or mitigate that risk before implementing the project.

Example of Cost-Benefit Analysis

To illustrate how a cost-benefit analysis is performed, let's consider a hypothetical example of an organization that is considering implementing a new inventory management system. Suppose the organization has identified the following costs and benefits associated with the project:

Costs:

Software cost: $20,000

Implementation cost: $5,000

Training personnel cost: $2,000

Additional operating costs: $1,000 per month

Benefits:

Inventory cost reduction: $3,000 per month

Increased sales: $2,000 per month

Inventory error reduction: $1,000 per month

Improved customer satisfaction: not quantifiable

Using this information, we can carry out a cost-benefit analysis in the following way:

Identification of costs and benefits: The identified costs include the cost of the software, the cost of implementation, the cost of training personnel, and additional operating costs. The identified benefits include a reduction in inventory costs, an increase in sales, and a reduction in inventory errors. An improvement in customer satisfaction is also identified, although it cannot be quantified.

Establishment of a baseline: The baseline is established by identifying the costs and benefits that are expected without the implementation of the project. Suppose that, without the new inventory management system, the organization expects to have inventory costs of $10,000 per month, sales of $20,000 per month, and inventory errors of $2,000 per month.

Identification of actual costs and benefits: Once the new inventory management system is implemented, the actual reduction in inventory costs, actual increase in sales, and actual reduction in inventory errors can be measured. Suppose these benefits turn out to be $4,000 per month, $2,500 per month, and $1,500 per month, respectively. Additionally, the additional operating costs turn out to be $1,500 per month.

Cost-benefit analysis: Using this information, we can compare the expected costs and benefits with the baseline and the actual costs and benefits to determine if the project is economically viable. The cost-benefit analysis is done in the following way:

Expected costs: $20,000 + $5,000 + $2,000 + ($1,000 x 12) = $39,000

Expected benefits: $3,000 + $2,000 + $1,000 = $6,000

Actual costs: $39,000 + ($1,500 x 12) = $57,000

Actual benefits: $4,000 + $2,500 + $1,500 = $8,000

Based on this analysis, we can see that the expected benefits exceed the expected costs, indicating that the project is economically viable. Furthermore, the actual benefits also exceed the actual costs, suggesting that the implementation of the new inventory management system has been a success.

Limitations of cost-benefit analysis

Although cost-benefit analysis is a useful tool for project evaluation, there are some important limitations that must be taken into account:

Difficulty in quantifying certain benefits: It can be difficult to quantify certain benefits, such as improvement in customer satisfaction. This can make it more difficult to determine if a project is economically viable.

Inaccurate assumptions: Cost-benefit analysis is based on assumptions, and if these assumptions are inaccurate, then the analysis can be incorrect. Therefore, it is important to ensure that the assumptions used are accurate and realistic.

Does not account for long-term costs and benefits: Cost-benefit analysis focuses on short-term costs and benefits and does not account for long-term costs and benefits. Therefore, a project may appear economically viable in the short term but not in the long term.

Conclusion

In summary, cost-benefit analysis is an important tool for project evaluation. It allows managers and decision-makers to determine whether a project is economically viable by comparing expected costs and benefits with actual costs and benefits. However, it is important to consider the limitations of cost-benefit analysis, such as the difficulty in quantifying certain benefits and inaccurate assumptions.

In addition to cost-benefit analysis, there are other tools and techniques that can be used for project evaluation, such as cost-effectiveness analysis, cost-utility analysis, and environmental and social impact analysis. Each of these tools and techniques has its own advantages and limitations, and the choice of the appropriate tool will depend on the project objectives and specific circumstances.

Ultimately, project evaluation is essential to ensure that organizations invest their resources effectively and efficiently. By evaluating the costs and benefits of a project, managers and decision-makers can make informed decisions about whether to continue with the project or not. This can help ensure that limited resources are used effectively to achieve the organization's goals and maximize its impact.

FINANCIAL ANALYSIS AND PROFITABILITY IN INDUSTRIAL ENGINEERING

Financial analysis and profitability analysis are fundamental tools in industrial engineering, as they allow for the evaluation of project viability and profitability, as well as strategic and tactical decision-making within a company. In this chapter, we will discuss the key concepts and techniques used in financial analysis and profitability, as well as their importance in industrial engineering.

Fundamental concepts

Before delving into financial analysis and profitability analysis, it is important to understand some fundamental concepts, such as the cost of capital, cash flow, discount rate, and break-even point.

The cost of capital refers to the rate of return required by investors to finance a project. This rate may be comprised of different components, such as the cost of debt, the cost of equity, and the cost of assets.

Cash flow refers to the money that enters and leaves a company during a given period. It is important to distinguish cash flow from accounting profits, as accounting profits do not always translate to cash flow.

The discount rate is the interest rate used to calculate the present value of future cash flows. This rate is used to evaluate the profitability of a project.

The break-even point refers to the sales level at which revenue equals costs. This point is important because it indicates the minimum sales level needed to avoid losses.

Financial analysis

Financial analysis refers to the study of a company's financial statements, such as the balance sheet, income statement, and cash flow statement. These financial statements provide important information about a company's financial position and performance.

The balance sheet shows the assets, liabilities, and equity of a company at a specific point in time. Assets are the resources that the company possesses, such as cash, accounts receivable, inventory, and property. Liabilities are the company's obligations, such as accounts payable and loans. Equity is the shareholders' investment in the company.

The income statement shows a company's revenues and expenses during a specific period. Revenues include sales of products or services, while expenses include production costs, administrative expenses, and taxes.

The cash flow statement shows a company's cash flows during a specific period. This financial statement is important because it shows the actual cash flow of the company, which allows for evaluating its ability to finance its operations and investments.

Different financial ratios, such as liquidity, profitability, and leverage, are used to analyze a company's financial statements.

Liquidity refers to a company's ability to pay its short-term debts. Different liquidity ratios are used, such as the current ratio and the quick ratio.

Profitability refers to a company's ability to generate profits from its operations. Different profitability ratios are used, such as return on equity, return on assets, and profit margin.

Leverage refers to a company's level of debt in relation to its assets and equity. Different leverage ratios are used, such as the debt ratio and the interest coverage ratio.

It is important to note that financial ratios should not be evaluated in isolation but should be analyzed in conjunction with the company's goals and financial situation.

Profitability analysis

Profitability analysis refers to evaluating the profitability of a project or investment. Different techniques are used in this type of analysis, such as net present value (NPV), internal rate of return (IRR), and payback period.

Net present value is a technique used to evaluate the profitability of an investment over time. NPV is calculated by subtracting the investment cost from the future cash flows discounted at an appropriate discount rate. If NPV is positive, the investment is profitable.

Internal rate of return is another technique used to evaluate the profitability of an investment. IRR is the discount rate at which NPV is equal to zero. If IRR is greater than the required rate of return, the investment is profitable.

Payback period refers to the time required to recover the initial investment. It is calculated by dividing the investment cost by the expected annual cash flow. If the payback period is less than the projected project time, the investment is profitable.

It is important to note that profitability analysis should consider not only expected cash flows but also investment risks, such as uncertainty in revenues and costs and changes in market conditions.

Application in Industrial Engineering

Financial and profitability analysis is a fundamental tool in industrial engineering, as it allows for the evaluation of project feasibility and profitability, as well as strategic and tactical decision making within a company.

In industrial engineering, financial and profitability analysis is applied in different areas, such as investment project evaluation, cost management, and financial decision making.

In investment project evaluation, profitability analysis is used to assess the feasibility of projects and determine if they are profitable or not. Additionally, financial analysis is used to identify the costs and revenues associated with projects and evaluate the sensitivity of results to different assumptions.

In cost management, financial analysis is used to identify the fixed and variable costs of a company and evaluate the profitability of its products and services. Different techniques, such as cost-volume-profit analysis, are also used to evaluate the impact of changes in costs and prices on the company's profitability.

In financial decision making, financial and profitability analysis is used to evaluate different options and determine the most profitable one. For example, different financing options, such as bond issuance or bank loans, can be evaluated to determine which option is the most profitable for the company in terms of costs and risks.

Furthermore, financial and profitability analysis is also used in evaluating the efficiency and effectiveness of a company's operations. For instance, different production options, such as outsourcing certain operations or automating processes, can be evaluated to determine the most profitable option in terms of costs and productivity.

In summary, financial and profitability analysis is a fundamental tool in industrial engineering, as it allows for the evaluation of project feasibility and profitability, as well as strategic and tactical decision making within a company. The application of financial and profitability analysis techniques enables companies to make informed decisions and minimize the risks associated with their operations and projects.

LEGAL ASPECTS AND REGULATIONS IN THE INDUSTRY

Industry is one of the main economic drivers of any country. However, the production of goods and services cannot be carried out without complying with a series of regulations and regulations established by government authorities and corresponding regulatory agencies. In this chapter, legal aspects and regulations applicable to the industry will be analyzed, in order to better understand the legal and regulatory framework that governs this sector.

Legal framework

The legal framework that regulates the industry varies from country to country and according to different legislations. However, there are certain general aspects that are common in most countries. One of the most important legal aspects is intellectual property. This includes patents, trademarks, copyrights, industrial designs, and models. Companies operating in the industry must comply with intellectual property laws, as this protects their property and allows them to compete fairly in the market.

Another important aspect of the legal framework is environmental protection. Companies must comply with environmental laws, which establish standards and regulations for waste production and management, gas emissions, and water pollution. Companies that do not comply with these regulations may face fines and sanctions, as well as civil lawsuits.

Safety regulations

Safety in the workplace is a critical aspect in the industry. The safety and health of workers must be protected by law, and companies must comply with workplace safety regulations. This includes implementing safety programs, training workers in safety, and maintaining equipment and machinery in safe conditions.

In addition, companies must also comply with safety regulations regarding the transport and handling of hazardous materials. This includes the transportation of hazardous chemicals, handling of explosive materials, and disposal of toxic waste. Companies that do not comply with these regulations may face criminal and civil penalties.

Employment regulations

Companies operating in the industry must also comply with labor and employment laws. This includes the payment of fair wages, protection of workers' rights, and compliance with workplace safety and health regulations. Companies must also comply with laws against discrimination and harassment in the workplace.

In addition, companies must comply with employment regulations regarding the hiring and firing of workers. This includes compliance with child labor regulations and protection of the rights of migrant workers.

Trade regulations

Companies operating in the industry must also comply with trade regulations and antitrust laws. Trade laws establish rules for international trade, including standards and regulations for the import and export of goods. Companies must also comply with antitrust laws, which seek to prevent the formation of monopolies and promote fair competition in the market. Companies that do not comply with these regulations may face sanctions and fines.

Quality regulations

The quality of products and services is another important aspect in the industry. Companies must comply with quality regulations established by government authorities and corresponding regulatory bodies. This includes food safety regulations, which establish standards and regulations for the production and handling of food and beverages. Companies must also comply with quality regulations regarding the production and supply of pharmaceuticals and medical devices.

Tax regulations

Companies operating in the industry must also comply with tax and fiscal laws. This includes the payment of taxes, fees, and tariffs, as well as compliance with accounting and auditing regulations. Companies must submit accurate and transparent financial reports and comply with accounting regulations established by government authorities.

Property and zoning regulations

Companies operating in the industry must also comply with property and zoning regulations. Property laws establish rules for the ownership and use of land and buildings. Companies must comply with zoning regulations, which establish standards and regulations for the use of land and construction of buildings. Companies must obtain necessary permits before constructing or modifying a building, and comply with construction and safety regulations.

Conclusion

In summary, the industry is subject to a wide range of regulations and laws that seek to protect workers, the environment, and the general public, and promote fair competition in the market. Companies operating in the industry must comply with these regulations and laws in order to operate effectively and responsibly. It is important for companies to understand the legal and regulatory framework governing the industry in their country and ensure compliance with all applicable regulations and laws. In this way, companies can protect their intellectual property, keep their workers safe and healthy, produce high-quality products, and contribute positively to the economy and society as a whole.

SUSTAINABILITY AND SOCIAL RESPONSIBILITY IN INDUSTRIAL ENGINEERING

Sustainability and social responsibility are two key topics in modern industrial engineering. As natural resources become increasingly scarce and concerns over the environmental and social impact of human activities grow, industrial engineers are tasked with designing and operating production systems that are sustainable and socially responsible. In this chapter, we will explore the concepts of sustainability and social responsibility in industrial engineering, and discuss some of the tools and strategies that engineers can use to integrate these concepts into their work.

Sustainability in Industrial Engineering

Sustainability refers to the ability to meet present needs without compromising the ability of future generations to meet their own needs. In industrial engineering, sustainability involves designing and operating production systems in a way that efficiently utilizes resources and minimizes environmental impact. This involves adopting practices that reduce the use of non-renewable resources, decrease greenhouse gas emissions, reduce waste generation, and promote biodiversity conservation.

One of the most important tools that industrial engineers can use to improve the sustainability of production systems is life cycle assessment (LCA). LCA is a methodology that allows for the evaluation of the environmental impact of a product or system throughout its entire life cycle, from raw material extraction to final disposal. By using LCA, engineers can identify critical points in the life cycle of a product or system and take steps to reduce its environmental impact.

Another important tool is design for environment (DfE). DfE involves considering the environmental impacts of a product or system from the initial design stage and seeking opportunities to reduce its environmental impact throughout its entire life cycle. This may involve the use of more sustainable materials and technologies, reducing energy and resource use, and implementing recycling and waste management practices.

Social Responsibility in Industrial Engineering

Social responsibility refers to the obligation of companies and organizations to operate in an ethical and responsible manner with regard to society as a whole. In industrial engineering, social responsibility involves adopting practices that promote social justice, equal opportunities, and the well-being of the community at large.

One key area of social responsibility for industrial engineers is occupational safety and health management. Engineers must design and operate production systems that are safe and healthy for workers, and take steps to prevent accidents and injuries in the workplace. This may include implementing safety measures such as the use of personal protective equipment and safety training for workers.

Another key area of social responsibility is supply chain management. Engineers must work with suppliers and contractors to ensure that their practices are socially responsible and comply with ethical and legal standards. This may include promoting fair labor practices, eliminating child and forced labor, and adopting sustainable environmental management practices.

In addition, industrial engineers must also consider social responsibility in the design of products and systems. This involves considering how products and systems will affect end users and society as a whole, and seeking opportunities to improve quality of life and social equity. For example, engineers can design products that are accessible to people with disabilities or that promote gender equality.

Tools and Strategies for Sustainability and Social Responsibility in Industrial Engineering

To integrate sustainability and social responsibility into their work, industrial engineers can use a variety of tools and strategies. Some of the most common tools and strategies include:

Standards and norms: Engineers can use a variety of standards and norms to guide their practices and ensure sustainability and social responsibility in their projects. Examples include ISO 14001 (environmental management system), ISO 45001 (occupational health and safety management system), and SA8000 (social responsibility standard).

Certifications and labels: Engineers can work with certifications and labels to demonstrate that their products and systems meet environmental and social standards. Examples include the LEED (Leadership in Energy and Environmental Design) certification for green buildings, and fair trade labels for socially responsible manufacturing.

Social impact analysis: Engineers can use tools such as social impact analysis to evaluate the impact of their projects on society as a whole. This may involve assessing positive impacts such as job creation or access to basic services, as well as identifying negative impacts such as social exclusion or loss of cultural heritage.

Stakeholder engagement and dialogue: Engineers can involve stakeholders such as the local community, workers, and interest groups in the design and implementation of projects. This can help ensure that projects are socially responsible and meet the needs and concerns of all stakeholders.

Innovation and technology: Engineers can use innovation and technology to develop sustainable and socially responsible solutions. This may involve using clean and renewable technologies, implementing more efficient processes, and exploring new forms of design and production.

Conclusion

Sustainability and social responsibility are critical issues for modern industrial engineering. As the world faces increasingly complex environmental and social challenges, industrial engineers have a responsibility to design and operate production systems that are sustainable and socially responsible. This requires considering not only efficiency and economic profitability, but also the impact of their practices on people and the planet.

Industrial engineers have a wide variety of tools and strategies at their disposal to integrate sustainability and social responsibility into their work. These tools and strategies include standards and norms, certifications and labels, social impact analysis, stakeholder engagement and dialogue, and innovation and technology.

It is important for industrial engineers to work collaboratively with other professionals and stakeholders to ensure that their practices are truly sustainable and socially responsible. This may involve collaboration with environmental scientists, occupational health and safety experts, local community representatives, and other interest groups.

Sustainability and social responsibility in industrial engineering are complex and constantly evolving issues. Industrial engineers must stay informed about the latest developments in these areas and be willing to adapt their practices accordingly. Through a collaborative, sustainability- and socially-responsible approach, industrial engineers can play a crucial role in building a more just and sustainable world.

TRENDS AND FUTURE PERSPECTIVES IN INDUSTRY

Currently, the industry is undergoing constant change due to various factors such as technological innovation, globalization, digital transformation, sustainability, and circular economy, among others. These trends are impacting the way companies operate and how they relate to their customers and the environment. In this chapter, these trends will be analyzed and future perspectives of the industry will be explored.

Technological Innovation

Technological innovation is one of the main trends transforming the industry. The digitization of processes and automation of tasks are improving the efficiency and productivity of companies. In addition, the implementation of technologies such as artificial intelligence, the internet of things, virtual and augmented reality, and robotics are generating new business opportunities and improving the customer experience.

An example of how technological innovation is impacting the industry is the case of the automotive industry. The incorporation of technologies such as autonomous driving and connectivity are transforming the way people use vehicles and how they interact with them.

Globalization

Globalization is another trend that is transforming the industry. The opening of markets and internationalization of companies are generating new business opportunities and allowing access to new customers and suppliers. However, globalization also presents challenges such as global competition and the need to adapt to different cultures and regulations.

Globalization has allowed companies from various sectors to establish themselves in different countries and expand their presence in the global market. An example of this is the case of technology companies, which have been able to establish their presence in different countries and regions thanks to globalization and connectivity.

Digital transformation

Digital transformation is another trend that is impacting the industry. The digitization of processes and the implementation of technologies such as big data, artificial intelligence, and the cloud are allowing companies to improve efficiency and productivity, as well as offer new digital services and products.

Digital transformation is also enabling companies to improve the customer experience, thanks to the implementation of digital solutions that allow for more efficient communication and personalized interaction. An example of this is the case of e-commerce companies, which have improved the customer experience through the implementation of digital solutions that allow for a more personalized and efficient shopping experience.

Sustainability and circular economy

Sustainability and the circular economy are increasingly important trends in the industry. Sustainability involves the need to reduce the environmental impact of companies and their products, while the circular economy aims to reduce waste and resource use through the reuse and recycling of materials.

Sustainability and the circular economy are transforming the way companies operate and how they relate to the environment. An example of this is the case of fashion companies, which are implementing sustainability and circular economy strategies throughout the supply chain, from production to the sale and recycling of products. This involves the use of sustainable materials, waste reduction, and the implementation of recycling and reuse processes.

Future perspectives of the industry

As for the future perspectives of the industry, it is expected that the trends mentioned above will continue to transform the sector in the coming years. Below are some of the most relevant future perspectives that will be analyzed.

Artificial intelligence and automation

Artificial intelligence and automation are two trends that will continue to transform the industry in the future. It is expected that the implementation of these technologies will continue to improve the efficiency and productivity of companies, as well as enable the creation of new products and services.

In addition, artificial intelligence and automation also have the potential to generate new jobs and improve working conditions, by allowing workers to focus on more complex and creative tasks.

Circular economy and sustainability

The circular economy and sustainability will continue to be key trends in the industry in the future. It is expected that companies will continue to adopt sustainable practices and circular economy strategies, and that more and more consumers will demand sustainable products and services.

The implementation of circular economy and sustainability strategies can also be an opportunity for companies to differentiate themselves from the competition and improve their corporate reputation.

Emerging technologies

Emerging technologies such as virtual and augmented reality, blockchain, and quantum computing will also have an impact on the industry in the future. It is expected that these technologies will allow for the creation of new products and services and improve the customer experience.

In addition, the implementation of these technologies can be an opportunity for companies to innovate and differentiate themselves from the competition.

Digital transformation

Digital transformation will continue to be a relevant trend in the industry in the future. It is expected that companies will continue to digitize processes and adopt new technologies to improve efficiency and productivity.

In addition, digital transformation can also be an opportunity for companies to adapt to changes in consumer preferences and improve the customer experience.

Conclusions

In conclusion, the industry is constantly changing due to various trends such as technological innovation, globalization, digital transformation, sustainability, and the circular economy. These trends are transforming the way companies operate and how they relate to their customers and the environment.

In the future, it is expected that these trends will continue to transform the industry and new trends will emerge, such as artificial intelligence and automation, the circular economy and sustainability, emerging technologies, and digital transformation.

Therefore, it is important for companies to be prepared to adapt to these changes and take advantage of the opportunities that arise. Those that manage to adapt and adopt sustainable and innovative practices will be the ones that are most successful in the future of the industry.

INTRODUCTION TO INDUSTRIAL METHODOLOGIES

In today's world, where competition is increasingly intense and globalized, it is crucial for companies to continually improve their production processes and services to stay in the market. Industrial methodologies are a key tool for achieving this continuous improvement.

Industrial methodologies are techniques and tools used to enhance process quality, reduce costs, increase productivity and efficiency, and maximize customer satisfaction. These methodologies are applicable to any industry, from manufacturing to services, and can be employed at any stage of the product or service lifecycle.

An industrial methodology is a structured and systematic set of techniques, tools, and processes applied to enhance the effectiveness and efficiency of a production process or service. These methodologies enable effective resource management, identification of problems and improvement opportunities, and the implementation of solutions for process optimization.

Various industrial methodologies exist, each designed to address specific needs and objectives. Some of the most common ones are described below:

Lean Manufacturing: Focuses on waste elimination and production optimization to improve process efficiency and quality, reducing production time and costs.

Six Sigma: Concentrates on reducing variability and improving quality to achieve stable and predictable processes, eliminating defects and errors in the process.

Kaizen: Based on continuous improvement and waste elimination, aiming to foster a culture of constant improvement by identifying and eliminating sources of waste and improving process quality.

Total Quality Management (TQM): Centers on customer satisfaction and continuous quality improvement, striving for total quality across all areas of the company, from production to customer service.

Business Process Management (BPM): Concentrates on optimizing business processes to identify and eliminate bottlenecks and inefficiencies, improving process efficiency and quality.

Each industrial methodology has specific tools for implementation. For instance, lean manufacturing uses tools like value stream mapping and kanban, while Six Sigma employs tools like root cause analysis and experimental design.

Implementing an industrial methodology is not a simple process; it requires constant commitment and dedication from the entire organization. The company's leadership must clearly define objectives and establish a strategy to achieve them. Employee training in the chosen methodology's tools and techniques is crucial.

Implementing an industrial methodology not only allows for the optimization of production processes but also contributes to the motivation and satisfaction of workers by providing tools and techniques to enhance their performance and contribution to the company. Additionally, the application of these methodologies can contribute to cost reduction and quality improvement, significantly impacting the company's profitability.

Industrial methodologies also enable better risk management in production processes. By identifying and addressing improvement opportunities, the likelihood of errors and process failures is reduced, decreasing the risk of product and service rejections, diminished customer satisfaction, and revenue loss.

Moreover, industrial methodologies can contribute to the environmental and social sustainability of the company. By reducing waste and inefficiency in processes, the consumption of natural and energy resources decreases, positively impacting the environment. Furthermore, improved quality and customer satisfaction can enhance the company's reputation and social responsibility.

In summary, industrial methodologies are a key tool for achieving continuous improvement in a company's production processes and services. These methodologies enable resource optimization, cost reduction, increased productivity and customer satisfaction, risk management, environmental and social sustainability, and improved company profitability.

It is important to note that implementing an industrial methodology is not an isolated process; it requires constant commitment from the entire organization. The company's leadership must clearly define objectives and establish a strategy to achieve them, involving all levels of the organization in the process. Employee training in the tools and techniques of the chosen methodology is essential, as is the continuous measurement and monitoring of results to ensure continuous improvement.

HISTORY OF INDUSTRIAL METHODOLOGIES AND THEIR EVOLUTION

The evolution of industrial methodologies has been a constant throughout history. From the industrial revolution to the present day, companies have sought to improve their processes to make them more efficient and effective. This chapter will describe the evolution of industrial methodologies over time, from early artisanal production systems to modern business management techniques.

Artisanal Production

Before the industrial revolution, most products were crafted in an artisanal manner. Workers were responsible for every stage of the production process, from acquiring raw materials to selling the final product. This form of production was limited in terms of the quantity of products that could be manufactured, and the products varied in quality.

However, artisanal production also had its advantages. Workers had a deep knowledge of the production processes, allowing them to make adjustments and improvements on the fly. Additionally, products were unique and customized, adding value.

The Industrial Revolution and Mass Production

The industrial revolution brought about a radical change in how products were manufactured. Technological advances led to the creation of machines capable of performing repetitive tasks much faster and more efficiently than manual laborers.

This marked the beginning of mass production.

Mass production enabled the manufacturing of large quantities of products quickly and efficiently. Furthermore, the products had consistent quality, improving product reliability. However, mass production also had its drawbacks; products became identical, losing their unique and personalized characteristics. Moreover, the production pace was fast, making workers feel like mere cogs in the machine.

Taylorism and Scientific Management

Taylorism is a management methodology developed by Frederick Winslow Taylor in the early 20th century. This methodology was based on the observation and analysis of production processes to enhance efficiency and productivity.

Taylor believed that business management should be a science, and production processes could be studied objectively and analytically. To achieve this, he developed techniques and tools such as time and motion studies, task analysis, process standardization, and worker selection and training.

Taylorism had a significant impact on industry, improving efficiency and productivity in production processes. However, it also faced criticism for turning workers into mere robots, diminishing creativity, and initiative.

Fordism and Assembly Line Production

Fordism is a management methodology developed by Henry Ford in the 1910s. It was based on assembly line production, a system where production tasks were divided into specific and repetitive tasks performed by specialized workers. This approach achieved greater efficiency and reduced production costs.

Fordism had a profound impact on industry, especially in automobile production. Assembly line production allowed the manufacturing of large quantities of vehicles at a lower cost, making automobiles more accessible to the general population.

However, assembly line production had its disadvantages. Workers performed very specific and repetitive tasks, which could be monotonous and boring. Additionally, it required substantial investments in machinery and equipment, making companies highly dependent on market demand.

Total Quality Era and Continuous Improvement

In the 1950s, new business management methodologies emerged, focusing on continuous improvement and total quality. These methodologies were based on the idea that continuous improvement was essential for maintaining competitiveness in the market.

One of the significant methodologies from this period was the Toyota Production System, developed by the Japanese company Toyota. This system emphasized continuous improvement of production processes and waste elimination to enhance efficiency and product quality.

The Toyota Production System had a profound impact on the industry and became a reference for companies worldwide. It also laid the groundwork for the Lean methodology, which focused on waste elimination and continuous improvement.

Total quality was also an important methodology during this time. It emphasized that quality should be a priority in all aspects of the company, from acquiring raw materials to selling the final product. Techniques and tools such as statistical quality control and ISO certification were developed for this purpose.

The Digital Era and Modern Industry

Currently, we are experiencing a new industrial revolution known as Modern Industry. This revolution is based on the digitization and automation of production processes, aiming to improve efficiency and productivity.

Modern Industry relies on technologies such as the Internet of Things, artificial intelligence, and data analysis. These technologies enable connection and communication between different teams and systems, allowing greater coordination and efficiency in production processes.

Moreover, Modern Industry focuses on product customization. Thanks to digitization and automation, it is possible to manufacture customized products efficiently and cost-effectively.

Conclusions

The evolution of industrial methodologies has been a constant throughout history.

From artisanal production systems to Modern Industry, companies have sought to improve their processes to become more efficient and effective.

Each methodology described in this chapter has its advantages and disadvantages and has been implemented at different times in history based on the needs of companies and available technological advances.

Currently, Modern Industry is transforming the way products are produced and consumed, with increased efficiency and product customization. However, it also presents new challenges in terms of worker training and adaptation to technological changes.

It is essential to highlight that, beyond specific methodologies, continuous improvement and the pursuit of efficiency in processes have always been fundamental in the industry. This has allowed companies to adapt to market changes and remain competitive over time.

In conclusion, the history of industrial methodologies reflects the evolution of the industry and society as a whole. Each new methodology has been an advancement in terms of efficiency and productivity but has also posed new challenges, requiring adaptation from both companies and workers. The industry continues to advance, and new methodologies and technologies are likely to emerge in the future, but the need for continuous improvement and efficiency will remain a constant in industrial history.

BASIC CONCEPTS OF INDUSTRIAL METHODOLOGIES

Industrial methodologies are a set of techniques and tools used to improve production processes in companies and enhance their efficiency and productivity. These methodologies have been developed over the years to adapt to the needs of each company and sector, and they are based on principles such as continuous improvement, waste elimination, and resource optimization.

This chapter will present the basic concepts of the most common industrial methodologies, such as Lean Manufacturing, Six Sigma, Kaizen, and Total Quality Management (TQM). The fundamental principles of each methodology will be explained, and a comparison will be made to assist managers and professionals in choosing the most suitable methodology for their company.

Lean Manufacturing

Lean Manufacturing is a methodology that focuses on waste elimination and the optimization of production processes. It is based on the concept that any activity that does not add value to the product or service is waste and should be eliminated.

The Lean methodology concentrates on reducing the seven types of wastes: overproduction, waiting time, transportation, unnecessary processes, excessive inventory, unnecessary motion, and defects. To achieve this, tools such as value stream mapping, process standardization, just-in-time (JIT), Kanban system, and continuous improvement are used.

Value stream mapping is a tool that identifies activities that add value and those that do not in the production process. With this information, waste can be eliminated, and the production process can be optimized. Process standardization is used to ensure that all employees follow the same steps in the production process, reducing errors and variability.

Just-in-time (JIT) is a system used to minimize inventory and reduce storage costs. Instead of producing large quantities of products and storing them, only the necessary amount is produced when needed. The Kanban system is used to manage production and inventory, employing a card system to indicate when more units of a product need to be produced.

Continuous improvement is a process involving the constant identification and elimination of waste. This is achieved through the formation of continuous improvement teams, the implementation of a feedback system, and the involvement of all employees in process improvement.

Six Sigma

Six Sigma is a methodology that focuses on reducing variability and eliminating defects in production processes. The goal of Six Sigma is to achieve a process where the number of defects is less than 3.4 per million opportunities.

Six Sigma is based on the DMAIC model (Define, Measure, Analyze, Improve, Control), a systematic approach to process improvement. The first step is defining the problem and establishing project objectives. The next step is measuring process variability and collecting data to identify critical points. A detailed analysis of the data is then conducted to identify the root causes of problems. With this information, solutions can be designed to improve the process. Finally, controls are established to ensure that the process maintains the desired quality level.

In Six Sigma, tools such as the Ishikawa diagram, FMEA (Failure Mode and Effect Analysis), process capability, and correlation analysis are used to identify and resolve problems in the production process.

Kaizen

Kaizen is a methodology that focuses on continuous improvement and is based on the concept that any process can be improved. Kaizen concentrates on process improvement through small constant changes rather than large changes all at once.

In Kaizen, tools such as process flow analysis, standardized work, improvement teams, and waste elimination are used to enhance processes. Process flow analysis identifies activities that do not add value and eliminates them. Standardized work ensures that all employees follow the same steps in the production process.

Improvement teams are groups of employees that regularly meet to identify and solve problems in production processes. Waste elimination is a constant practice in Kaizen that focuses on removing any activity that does not add value.

Total Quality Management (TQM)

Total Quality Management (TQM) is a methodology that focuses on quality in all aspects of the company, not just in production. TQM is based on the concept that quality is the responsibility of all employees, not just production workers.

In TQM, tools such as quality planning, quality control, continuous improvement, and customer satisfaction are used to improve quality throughout the company. Quality planning focuses on establishing quality objectives and developing a plan to achieve them. Quality control is used to ensure that products or services meet established quality standards.

Continuous improvement focuses on identifying and eliminating waste in all aspects of the company. Customer satisfaction focuses on ensuring that products or services meet customer expectations.

Comparison of Methodologies

Each of the mentioned industrial methodologies has its strengths and weaknesses. Lean Manufacturing focuses on waste elimination and process optimization but may not be suitable for companies producing highly customized products. Six Sigma focuses on reducing variability and eliminating defects but may be expensive to implement.

Kaizen focuses on continuous improvement through small constant changes, making it suitable for companies seeking constant improvement rather than a one-time change. TQM focuses on quality in all aspects of the company, making it suitable for companies looking to improve quality in all aspects of their operation.

In terms of implementation, Lean Manufacturing and Kaizen are relatively easy to implement as they focus on small changes and constant improvements in the production process. Six Sigma and TQM, on the other hand, require a greater investment in time and resources for implementation.

Regarding results, Lean Manufacturing and Kaizen tend to produce quick and tangible results in terms of waste reduction and efficiency improvement. Six Sigma and TQM may take more time to produce tangible results but can lead to significant improvements in product or service quality.

It is important to note that none of these methodologies is a one-size-fits-all solution for all companies and situations. Each company should analyze its specific needs and choose the methodology that best suits those needs.

Conclusions

In summary, industrial methodologies are systematic and structured approaches used to improve efficiency and quality in production. There are various industrial methodologies, including Lean Manufacturing, Six Sigma, Kaizen, and TQM, each with its strengths and weaknesses.

The goal of Lean Manufacturing is to eliminate waste in the production process and optimize processes. Six Sigma focuses on reducing variability and eliminating defects in the production process. Kaizen focuses on continuous improvement through small constant changes, while TQM focuses on quality in all aspects of the company.

Each company should analyze its specific needs and choose the methodology that best suits those needs. It is also important to consider that implementing these methodologies requires a significant investment in time and resources, and results may take time to materialize. However, once implemented, these methodologies can lead to significant improvements in production efficiency and quality.

IMPORTANCE OF INDUSTRIAL METHODOLOGIES IN THE CONTINUOUS IMPROVEMENT OF PROCESSES

The deafening roar of machines filled the production floor as workers moved with skill, performing their daily tasks in a well-coordinated dance. However, despite the constant hum of industrial activity, something did not seem to be working quite right.

Such was the case for ABC Company, an electronics manufacturing factory that had been experiencing issues with its production line. Delivery delays and inconsistent product quality had led the company to lose customers and suffer a decline in revenue. In search of a solution, ABC's management team decided to implement industrial methodologies into their production process.

Continuous process improvement through industrial methodologies has been a valuable tool for companies worldwide. These methodologies enable companies to optimize their production processes and improve efficiency in their supply chain, resulting in increased customer satisfaction and higher profits. In this chapter, we will explore the importance of industrial methodologies in continuous process improvement and how they can be successfully implemented in a company.

What are industrial methodologies?

Industrial methodologies are a set of techniques and tools used to improve production processes within a company. These methodologies are designed to help companies optimize their operations and improve the quality of their products, leading to higher customer satisfaction and increased profits.

There are several popular industrial methodologies, each with its strengths and weaknesses. Some common industrial methodologies include:

Lean Manufacturing: Focuses on eliminating any activity that does not add value to the production process, allowing companies to reduce costs and improve product quality.

Six Sigma: Concentrates on reducing variability in production processes, helping companies enhance product quality and reduce costs associated with defects.

Theory of Constraints: Focuses on identifying bottlenecks in the production process and implementing solutions to eliminate them, allowing companies to improve efficiency in their supply chain.

Total Productive Maintenance: Concentrates on improving the maintenance of production equipment, enabling companies to reduce downtime and enhance efficiency in their production process.

5S: Focuses on organizing and cleaning the workspace, helping companies improve workplace safety and efficiency in the production process.

The choice of the appropriate industrial methodology depends on the specific goals of the company and the areas it wants to improve in its production process.

The importance of industrial methodologies in continuous process improvement

Industrial methodologies are essential for continuous process improvement in a company. By implementing these methodologies, companies can improve the quality of their products, reduce costs, and enhance efficiency in their supply chain. This, in turn, can lead to higher customer satisfaction and increased profits.

Additionally, industrial methodologies also help companies identify and resolve issues in their production processes more quickly and efficiently. This is particularly important in an increasingly competitive business environment where speed and efficiency are key to success.

Another benefit of industrial methodologies is that they promote collaboration and teamwork among company employees. By working together to identify and solve problems in the production process, employees can develop a deeper understanding of processes and improve their ability to work together effectively.

How to implement industrial methodologies in a company

Implementing industrial methodologies in a company can be a complex process, but there are some general guidelines that companies can follow to ensure the success of their implementation:

Identify Objectives: Clearly identify the goals the company aims to achieve by implementing an industrial methodology. Is it looking to reduce costs, improve product quality, or enhance efficiency in the supply chain? Once the objectives are identified, it becomes easier to select the appropriate industrial methodology to achieve them.

Training: Ensure that company employees receive proper training on the industrial methodology being implemented. This allows them to understand the methodology, its benefits, and how they can contribute to its success.

Communication: Clearly communicate the implementation of the industrial methodology to all employees. This helps them understand why the methodology is being implemented, what is expected of them, and how they can contribute to its success.

Problem Identification and Resolution: Identify problems in the production process and resolve them before implementing the industrial methodology. This ensures that the implementation of the methodology is more effective.

Monitoring and Evaluation: Monitor and evaluate the implementation of the industrial methodology to ensure that desired objectives are being achieved. This allows the company to make adjustments if necessary and ensure that the methodology is being implemented effectively.

Conclusion

Industrial methodologies are essential for continuous process improvement in a company. By implementing these methodologies, companies can improve product quality, reduce costs, and enhance efficiency in their supply chain. Moreover, industrial methodologies help companies identify and resolve issues in their production processes more quickly and efficiently. By following some general guidelines, companies can successfully implement industrial methodologies and improve their production process effectively.

TYPES OF INDUSTRIAL METHODOLOGIES AND THEIR APPLICATIONS

The industry has evolved significantly in recent years, giving rise to a variety of methodologies aimed at improving the efficiency and quality of industrial processes. In this chapter, we will explore some of the most widely used methodologies in the industry, along with their applications and benefits.

Industrial Methodologies:

Lean Manufacturing:

Lean Manufacturing is a methodology that focuses on reducing production times and eliminating waste in the process. It is based on eliminating any activity that does not add value to the final product or service. Some of the tools used in this methodology include Kanban, Just-In-Time (JIT) production, and continuous improvement.

One of the most well-known applications of Lean Manufacturing is in automobile manufacturing. Toyota was one of the first companies to adopt this methodology, enabling them to improve the quality and efficiency of their production. Today, many companies use this methodology to optimize their processes and improve profitability.

Six Sigma:

Six Sigma is another methodology that seeks to improve the quality of industrial processes. It focuses on eliminating defects in processes and reducing variability. The goal of Six Sigma is to achieve nearly perfect quality in production.

Six Sigma uses a statistical methodology to measure quality and reduce defects in processes. It is divided into five phases: Define, Measure, Analyze, Improve, and Control (DMAIC). Some tools used in Six Sigma include control charts, Pareto analysis, and linear regression.

Six Sigma has been successfully applied in the food industry, medical device manufacturing, and electronic component production. Companies like General Electric and Motorola have implemented this methodology in their processes, improving product quality and reducing production costs.

Total Productive Maintenance (TPM):

Total Productive Maintenance focuses on improving the efficiency of industrial equipment. It is based on preventing equipment failures and continuously improving them to reduce downtime and enhance productivity.

TPM consists of eight pillars: Focused improvement, Autonomous maintenance, Planned maintenance, Training and education, Quality maintenance, Safety maintenance, Administrative maintenance, and Continuous improvement. Each of these pillars focuses on a specific area of equipment maintenance.

TPM has been successfully used in manufacturing, food production, and the chemical industry. Companies like Coca-Cola and Toyota have implemented TPM in their processes, improving equipment efficiency and reducing maintenance costs.

Quick Response Manufacturing (QRM):

Quick Response Manufacturing focuses on reducing production response times and eliminating wait times in the process. It is based on eliminating production bottlenecks and reducing tool change times. Tools used in QRM include value stream analysis and capacity management.

QRM has been successfully used in manufacturing, food production, and the healthcare industry. Companies like Harley-Davidson and L'Oréal have implemented QRM in their processes, reducing product delivery times and improving customer satisfaction.

Design for Six Sigma (DFSS):

Design for Six Sigma focuses on improving the quality of product design by identifying and eliminating design defects. Tools used in DFSS include risk analysis and customer voice evaluation.

DFSS has been successfully used in the automotive industry, medical device production, and the electronics industry. Companies like Ford and Philips have implemented DFSS in their design processes, improving product quality and reducing production costs.

Theory of Constraints (TOC):

The Theory of Constraints focuses on identifying and eliminating production bottlenecks. It involves identifying the most critical constraint in the process and eliminating impediments that limit production. Tools used in TOC include critical chain analysis and inventory management.

TOC has been successfully used in manufacturing, food production, and the healthcare industry. Companies like Procter & Gamble and Eli Lilly have implemented TOC in their processes, identifying production bottlenecks and improving process efficiency.

Conclusion:

The implementation of these methodologies in the industry can lead to a significant improvement in process efficiency and quality, resulting in increased profitability and customer satisfaction. Each methodology focuses on a specific area of improvement, but they all share the common goal of pursuing excellence in production.

It is important to note that there is no perfect methodology, and each company must evaluate which is most suitable for its processes and objectives. Implementing these methodologies requires a change in corporate culture, investment in training, and active participation from all employees.

Ultimately, the application of these methodologies in the industry can be the key to improving process efficiency and quality, translating into greater profitability and customer satisfaction. It is crucial for companies to be willing to invest in the implementation of these methodologies and in the training of their personnel to ensure long-term success.

PROCESSES AND PROCEDURES OF INDUSTRIAL METHODOLOGIES

In the industrial world, efficiency and productivity are fundamental to achieving success. To achieve this, various methodologies have been developed to improve the processes and procedures of companies, with the aim of optimizing the use of resources and reducing costs. In this chapter, we will explore the main industrial methodologies and how they can be applied in different business contexts.

Optimizing processes and procedures can be a complex challenge for companies. In a constantly changing environment, it is important to have tools and strategies that allow quick adaptation to new market demands and maintain a competitive position. Below, we will look at the most popular methodologies and how they are applied in the industry.

Lean Manufacturing Methodology

The Lean Manufacturing methodology, also known as lean production, is a technique that focuses on eliminating anything that does not add value to the production process. This technique originated in Japan in the 1950s, thanks to Toyota and its production system. The main goal of this methodology is to reduce costs, increase efficiency, and improve product quality.

The Lean Manufacturing methodology is based on five fundamental principles:

Identify value: The first step is to determine the value offered to the customer and how this value can be improved.

Map the value stream: Once the value is identified, the production process should be analyzed to identify activities that add value and those that do not.

Create a continuous flow: The goal is to eliminate activities that do not add value and create a continuous workflow.

Establish pull production: Instead of producing based on demand, production is based on actual consumption, avoiding inventory accumulation.

Strive for perfection: The ultimate goal is the continuous improvement of the process by eliminating activities that do not add value and improving those that do.

The application of the Lean Manufacturing methodology can be beneficial for companies, as it allows them to reduce costs, improve efficiency, and increase product quality. Additionally, this methodology can be applied in any type of company, regardless of its size or sector.

Six Sigma Methodology

The Six Sigma methodology is a technique that aims to improve the quality of production processes and reduce errors or defects in products. This methodology is based on a systematic and rigorous approach to identify and correct problems in the production process.

The main goal of the Six Sigma methodology is to reduce variation in the production process, thereby achieving a more consistent and higher-quality final product. To achieve this, two quality levels are established:

Sigma 6: The goal is to reduce defects by 99.99966%, translating to only 3.4 defects per million products.

Sigma 3: The goal is to reduce defects by 93.32%, translating to 66.800 defects per million products.

The Six Sigma methodology is based on five phases:

Define: In this phase, objectives and project scopes are established, customers are identified, and process requirements are set.

Measure: In this phase, the variation of the production process is measured, and quality indicators are established.

Analyze: In this phase, collected data is analyzed, and causes of problems or defects are identified.

Improve: In this phase, solutions are implemented to improve the production process, and tests are conducted to assess their effectiveness.

Control: In this phase, measures are established to maintain implemented changes, and the production process is monitored to ensure it stays within established quality standards.

The Six Sigma methodology can be beneficial for companies as it allows them to reduce errors and defects in products, improve quality, and reduce costs. Additionally, it focuses on the customer and satisfying their needs, which can enhance the company's reputation and increase customer loyalty.

5S Methodology

The 5S methodology is a technique that focuses on the organization and cleanliness of workspaces. This technique was developed in Japan by Toyota as part of its production system.

The 5S methodology is based on five principles:

Sort: Identify and separate necessary elements from unnecessary ones in the workspace.

Set in order: Once necessary elements are identified, organize them so they are easy to find and use.

Shine: Maintain a clean and organized workspace to avoid the accumulation of waste and dirt.

Standardize: Establish procedures and standards to consistently maintain a clean and organized workspace.

Sustain: Consistently maintain a clean and organized workspace by implementing procedures and standards.

The 5S methodology can be beneficial for companies as it improves the organization and cleanliness of the workspace, potentially increasing efficiency and productivity. Additionally, it can enhance workplace safety by reducing the risks of accidents and injuries.

Conclusion

In conclusion, industrial methodologies are valuable tools for improving the processes and procedures of companies. The Lean Manufacturing methodology focuses on eliminating activities that do not add value and creating a continuous workflow, reducing costs, and improving efficiency. The Six Sigma methodology focuses on improving quality and reducing errors and defects in products, increasing customer satisfaction, and enhancing the company's reputation. The 5S methodology focuses on organizing and cleaning the workspace, potentially increasing efficiency and workplace safety. Each of these methodologies can be applied in different business contexts, regardless of the size or sector of the company. It's important to remember that the implementation of these methodologies requires commitment and dedication from both the company and its employees, but the benefits can be significant in terms of efficiency, quality, and profitability.

It is recommended that companies assess their specific needs and objectives before choosing an industrial methodology to implement. Additionally, providing training and support to employees is crucial for successful implementation.

Lastly, it is important to remember that industrial methodologies are not magical solutions to solve all of a company's problems. These tools are useful but must be implemented strategically and adapted to the specific needs and objectives of the company. The key to success lies in constant commitment and dedication to continuous improvement.

In summary, industrial methodologies are a set of techniques and tools used to enhance the processes and procedures of businesses. The Lean Manufacturing methodology concentrates on eliminating non-value-adding activities and establishing a continuous workflow. The Six Sigma methodology centers on enhancing quality, reducing errors and defects in products, and aiming for customer satisfaction. The 5S methodology emphasizes organizing and cleaning the workspace, potentially improving efficiency and workplace safety. Each of these methodologies is applicable across various business domains and can contribute to enhancing efficiency, quality, and profitability. The successful implementation of these methodologies requires commitment and ongoing dedication from both the company and its employees, tailored to the specific needs and goals of the enterprise.

TOOLS AND TECHNIQUES OF INDUSTRIAL METHODOLOGIES

Industrial methodologies are tools and techniques that companies can use to improve the quality, efficiency, and profitability of their production processes. These methodologies are based on the identification and elimination of waste, problems, and errors in production processes.

Industrial methodologies consist of a set of tools and techniques that companies can use to enhance their production processes and increase profitability. Some of the most common industrial methodologies include Lean Manufacturing, Six Sigma, Total Quality Management, Just-in-Time (JIT), Poka-yoke, and Kaizen. Each methodology focuses on specific aspects of production, but they all share the common goal of improving the quality, efficiency, and profitability of processes.

Lean Manufacturing

Lean Manufacturing is a methodology that focuses on eliminating waste in production. Waste includes anything that does not add value to the product or the production process. Waste can include wait times, unnecessary movement, overproduction, defects, excess inventory, among others.

The Lean Manufacturing process involves identifying and eliminating waste in production. Workers use tools and techniques such as Value Stream Mapping and Process Flow Analysis to identify waste in production and design more efficient processes.

Lean Manufacturing is especially useful for companies that produce large quantities of similar products. By eliminating waste in production, companies can reduce costs and improve process efficiency.

Six Sigma

Six Sigma is a methodology used to improve the quality of products and production processes. It is based on the idea that errors and problems in production processes can be identified and eliminated through data analysis and the implementation of solutions.

The Six Sigma process involves the definition, measurement, analysis, improvement, and control (DMAIC) of production processes. Workers use tools and techniques such as Root Cause Analysis and Design of Experiments to identify problems in production and design solutions to improve the quality and efficiency of processes.

Six Sigma is especially useful for companies that produce high-quality products requiring a precise and well-defined production process. By improving product and production process quality, companies can increase customer satisfaction and enhance their market reputation.

Total Quality Management (TQM)

Total Quality Management (TQM) is a methodology used to improve the quality of products and production processes by implementing a quality focus throughout the entire company. It is based on the idea that quality is the responsibility of all workers and departments in the company.

The TQM process involves identifying and eliminating problems and waste in all company processes. Workers use tools and techniques such as Pareto Analysis and Ishikawa Diagram to identify problems and design solutions to improve quality and process efficiency.

TQM is especially useful for companies aiming to improve the quality of their products and production processes across all departments and processes. By implementing a quality focus throughout the company, companies can improve customer satisfaction and the company's reputation.

Just-in-Time (JIT)

Just-in-Time (JIT) is a methodology used to improve the efficiency of production processes by delivering materials, parts, and components just when they are needed for production. It is based on the idea that excess inventory and overproduction are wastes that can be eliminated through just-in-time delivery.

The JIT process involves identifying and eliminating waste in production through just-in-time delivery of materials, parts, and components. Workers use tools and techniques such as Kanban and Continuous Flow to design more efficient processes and reduce wait times and production costs.

JIT is especially useful for companies producing products with variable demand. By reducing excess inventory and overproduction, companies can lower costs and improve production process efficiency.

Poka-yoke

Poka-yoke is a methodology used to prevent errors and problems in production by implementing error prevention devices and systems. It is based on the idea that errors in production are inevitable but can be prevented through the implementation of error prevention devices and systems.

The Poka-yoke process involves identifying errors and problems in production and designing and implementing error prevention devices and systems. Workers use tools and techniques such as Poka-yoke Design to design more efficient processes and reduce errors and problems in production.

Poka-yoke is especially useful for companies producing complex and high-quality products requiring a precise and well-defined production process. By preventing errors and problems in production, companies can improve product quality and reduce production costs.

Kaizen

Kaizen is a methodology used to achieve continuous improvement in production processes through the implementation of small improvements in processes. It is based on the idea that continuous improvement is a constant process, and small improvements in production processes can add up to significant improvements over time.

The Kaizen process involves identifying and eliminating waste and problems in production through the implementation of small improvements in processes. Workers use tools and techniques such as Kaizen Blitz and Value Analysis to design more efficient processes and reduce waste and problems in production.

Kaizen is especially useful for companies seeking continuous improvement in their production processes over time. By implementing small improvements in production processes, companies can achieve significant improvements in the quality, efficiency, and profitability of their production processes.

Conclusion

In summary, industrial methodologies are tools and techniques that companies can use to improve the quality, efficiency, and profitability of their production processes. The implementation of these methodologies can help companies reduce production costs, improve product quality, and increase customer satisfaction.

The six methodologies described in this chapter are just some of the many tools and techniques available for improving production processes in companies. Each company should assess its own needs and goals and select the tools and techniques that best suit its requirements.

It is essential to note that the implementation of these methodologies is not a one-time process but should be a continuous improvement process. Companies must be willing to adapt and change their production processes as business needs and goals evolve.

Furthermore, the successful implementation of these methodologies requires commitment from the company's management and collaboration from all employees. All team members must be willing to learn new tools and techniques and work together to implement changes in production processes.

Ultimately, the implementation of these industrial methodologies can help companies improve the quality of their products, reduce production costs, and enhance the efficiency of their production processes. This can lead to increased customer satisfaction, a better company reputation, and long-term business profitability.

CHOOSING THE APPROPRIATE INDUSTRIAL METHODOLOGY FOR EACH PROJECT

Industrial methodology is a set of techniques, tools, and processes used to design, develop, and improve industrial products and processes. The selection of the appropriate methodology is essential for the project's success, as it can impact process efficiency, costs, and the quality of the final product.

In this chapter, five common industrial methodologies will be described: Design for Manufacturability (DFM), Concurrent Engineering, Design for Quality (DFQ), Lean methodology, and Six Sigma methodology. Additionally, examples of how these methodologies can be applied in different industrial projects will be provided.

Design for Manufacturability (DFM)

Design for Manufacturability (DFM) is a methodology that focuses on designing products to be easy and cost-effective to manufacture. This methodology is used to reduce production costs and improve the quality of the final product.

An example of how DFM methodology can be applied is in the production of plastic parts. In this project, DFM methodology can be utilized to design parts that are easy to mold, reducing production time and costs. For instance, the part can be designed in a way that requires a minimal amount of materials and is easy to extract from the mold without causing damage.

Concurrent Engineering

Concurrent Engineering is a methodology used to expedite product development by involving all stakeholders in the design process from the outset. This methodology is employed to enhance process efficiency and reduce development time.

An example of applying Concurrent Engineering methodology is in the design of a new electronic product. In this project, all stakeholders, including engineers, designers, suppliers, and end-users, can be involved in the design process from the beginning. By working together, stakeholders can quickly identify and address issues, reducing development time and improving the quality of the final product.

Design for Quality (DFQ)

Design for Quality (DFQ) is a methodology used to design high-quality products from the outset. This methodology is employed to enhance the quality of the final product and reduce production costs and cycle times.

An example of applying DFQ methodology is in the production of automobiles. In this project, DFQ methodology can be used to design cars that are safe, reliable, and of high quality. Design for Quality can help identify potential issues in the design and prevent quality defects during the production phase. For example, risk analysis tools can be used to identify and mitigate safety risks in the car design.

Lean Methodology

Lean methodology is an approach that focuses on waste elimination and continuous improvement. This methodology is suitable for projects aiming to reduce costs, increase productivity, and improve the quality of the final product.

An example of applying Lean methodology is in food production. In this project, processes that do not add value, such as wait times, unnecessary movements, and material transport, can be identified. Lean techniques like Kaizen (continuous improvement), Just-in-Time (lean production), and Value Stream Mapping can be applied to reduce costs, improve efficiency, and enhance the quality of the final product.

Six Sigma Methodology

Six Sigma methodology is a statistical approach to process improvement used to reduce defects and enhance the quality of the final product. This methodology relies on data collection and analysis to identify and eliminate root causes of issues in the process.

An example of applying Six Sigma methodology is in the production of medical devices. In this project, processes causing defects can be identified, and data can be collected to analyze process performance. Six Sigma methodology utilizes statistical tools to identify root causes of problems and improve the process to reduce defects and enhance the quality of the final product.

Selection of the Appropriate Methodology

Choosing the right methodology depends on the project type and business objectives. Below are some factors to consider when selecting an industrial methodology:

Project Type: The type of project can influence the choice of the appropriate methodology. For example, Lean methodology may be suitable for mass production projects, while Six Sigma methodology may be more suitable for continuous improvement projects.

Business Objectives: Business objectives can influence the choice of the appropriate methodology. If the business objective is to reduce costs, Lean methodology may be more suitable. If the objective is to improve quality, DFQ or Six Sigma methodologies may be more appropriate.

Available Resources: Available resources, including time, budget, and staff expertise, can influence the choice of the appropriate methodology. Some methodologies may require more resources than others, so it is important to consider resource availability before selecting a methodology.

Corporate Culture: Corporate culture can influence the choice of the appropriate methodology. For example, if the corporate culture values continuous improvement, Six Sigma methodology may be more suitable. If the culture values efficiency, Lean methodology may be more appropriate.

Conclusion

In conclusion, selecting the appropriate methodology is crucial for project success.

The right industrial methodology can enhance process efficiency, reduce costs, and improve the quality of the final product. When selecting an industrial methodology, it is important to consider project type, business objectives, available resources, and corporate culture. The methodologies described in this chapter, Design for Manufacturability (DFM), Concurrent Engineering, Design for Quality (DFQ), Lean methodology, and Six Sigma, are just a few of the many options available for selecting an appropriate methodology. A careful evaluation of each methodology is essential, considering that each project is unique and may require a combination of different methodologies for optimal results.

Additionally, it is important to note that the selected industrial methodology is not a one-size-fits-all solution. Each project is unique and may require a combination of different methodologies to achieve the best results.

Lastly, the successful implementation of an industrial methodology depends not only on choosing the right methodology but also on proper project planning, execution, and monitoring. Adequate staff training, allocation of appropriate resources, and continuous process evaluation are fundamental to project success.

In summary, selecting the right methodology is essential to improving process efficiency, reducing costs, and enhancing the quality of the final product in industrial projects. The choice of the right methodology depends on project type, business objectives, available resources, and corporate culture. When selecting a methodology, a careful evaluation is necessary, considering that each project is unique and may require a combination of different methodologies for optimal results. Additionally, proper project planning, execution, and monitoring are crucial for the successful implementation of the selected methodology.

IMPLEMENTATION OF INDUSTRIAL METHODOLOGIES IN THE COMPANY

The implementation of industrial methodologies in the company is a key process for improving efficiency and productivity in production. Industrial methodologies are techniques and tools used to enhance processes and reduce costs. Some of the most popular industrial methodologies include Lean Manufacturing, Six Sigma, Theory of Constraints, and Total Productive Maintenance (TPM).

To implement an industrial methodology in the company, several important steps must be followed. First, it is essential to identify the needs and goals of the company. Clear and specific objectives, such as reducing costs, improving product quality, and increasing production efficiency, should be established. Additionally, it is important for all employees to understand the goals of the methodology and how it will be applied in their daily work.

The next step in the implementation of industrial methodologies is to identify and analyze existing business processes. This includes identifying bottlenecks, points of inefficiency, and waste in production. Detailed analyses of each process should be conducted to identify areas for improvement.

Once issues in the processes have been identified, improvements should be designed and implemented. This may involve reorganizing processes, eliminating waste, automating processes, and improving quality. It is important to involve employees in the implementation of improvements to ensure success.

After implementing improvements, it is important to monitor and control processes to ensure that objectives are being achieved. Key performance indicators should be established, and periodic analyses should be conducted to assess the success of improvements. If issues are identified, adjustments and additional improvements should be made.

The implementation of industrial methodologies can be a challenging process, but it can be a powerful tool for improving efficiency and productivity. Careful evaluation of methodologies and selecting the one that best suits the needs and goals of the company is crucial. Employee training and education are key to ensuring the success of the implementation.

Below are some examples of companies that have successfully implemented industrial methodologies:

Toyota: Toyota is one of the most successful companies in implementing Lean Manufacturing. The company has used this methodology to improve production efficiency and reduce costs. Implementing Lean Manufacturing has allowed Toyota to produce more vehicles with fewer resources and improve the quality of its products. The company has also implemented Total Productive Maintenance (TPM) to enhance efficiency and quality in production.

General Electric: General Electric has implemented the Six Sigma methodology throughout its company. This methodology has enabled the company to improve the quality of its products and services, reduce costs, and increase efficiency in production. The implementation of Six Sigma has been key in transforming GE into a more quality- and efficiency-focused company. The company has also implemented the Design for Six Sigma (DFSS) methodology to improve the quality of product and service design.

Nestlé: Nestlé has implemented the Total Productive Maintenance (TPM) methodology in its production operations worldwide. This methodology has allowed the company to reduce costs and improve efficiency in production. Nestlé has also used Lean Manufacturing methodology to eliminate waste in production and improve the quality of its products.

Boeing: Boeing has implemented the Theory of Constraints methodology in its aircraft production. This methodology has enabled the company to identify and eliminate bottlenecks in production, improving efficiency and reducing costs.

Boeing has also used Lean Manufacturing methodology to enhance efficiency in production and reduce lead times.

Amazon: Amazon has implemented Lean Manufacturing methodology in its logistics and storage operations. The company has used this methodology to reduce lead times and improve efficiency in product delivery. Amazon has also used the Six Sigma methodology to improve the quality of its services and reduce errors in product delivery.

In summary, the implementation of industrial methodologies in the company is essential for improving efficiency, reducing costs, and enhancing the quality of products and services. The selection of the appropriate methodology should be based on the needs and goals of the company and should involve all employees to ensure success. Companies that have successfully implemented industrial methodologies have achieved significant improvements in efficiency, quality, and profitability.

EVALUATION AND MONITORING OF THE EFFECTIVENESS OF INDUSTRIAL METHODOLOGIES

The methodology review meetings are an effective tool for monitoring the effectiveness of industrial methodology. During these meetings, Key Performance Indicators (KPIs) are reviewed, and the obtained results are discussed. Additionally, improvement opportunities are identified, and action plans are established to implement changes.

It is important that these meetings are held consistently, involving all team members to ensure that objectives are being met and to foster collaboration in implementing changes.

During these meetings, it is necessary to review and analyze the established KPIs for evaluating the performance of industrial methodology. These KPIs may include aspects such as production time, product quality, machinery performance, and compliance with safety standards. Clear goals should be defined for each KPI.

Moreover, these meetings serve to identify improvement opportunities in processes and the implementation of industrial methodology. Encouraging ideas and suggestions from all team members is essential for identifying these opportunities and establishing action plans to implement changes.

In summary, methodology review meetings are an essential tool for monitoring the effectiveness of industrial methodologies. They should be conducted consistently, reviewing KPIs and identifying improvement opportunities to implement changes and achieve continuous improvement in company processes.

Internal Audits

Internal audits are a valuable tool for assessing compliance with quality standards and identifying improvement opportunities in industrial methodology. These audits are conducted regularly, focusing on evaluating process efficiency, product quality, and compliance with established standards.

Impartiality is crucial in conducting audits, involving all team members to ensure that industrial methodology objectives are met. During audits, all aspects of industrial methodology, from product quality to process efficiency, should be reviewed.

Clear objectives should be set for audits, and action plans should be established to implement changes. Constant monitoring is essential to ensure that changes are implemented and objectives are met.

In summary, internal audits are an essential tool for evaluating compliance with quality standards and identifying improvement opportunities in industrial methodology. They should be conducted regularly, involving all team members, and action plans should be established to implement changes and achieve continuous improvement in company processes.

Customer Satisfaction Surveys

Customer satisfaction surveys are a valuable tool for evaluating the effectiveness of industrial methodology from the customer's perspective. These surveys focus on measuring customer satisfaction with the company's products and services, as well as the efficiency of processes and meeting expectations.

Regularly conducting these surveys and involving all customers is important to ensure that their expectations are met and to identify improvement opportunities in industrial methodology. Clear objectives should be set for surveys, and action plans should be established to implement changes based on the results obtained.

Customer satisfaction surveys should include questions that assess aspects such as product quality, service efficiency, customer service, and meeting expectations. Specific questions should be established for evaluating each of these aspects, with the opportunity for additional comments to obtain detailed and specific information.

In summary, customer satisfaction surveys are an essential tool for evaluating the effectiveness of industrial methodology from the customer's perspective. They should be conducted regularly, involving all customers, and action plans should be established to implement changes based on the results obtained.

Data Analysis

Data analysis is an essential tool for monitoring the effectiveness of industrial methodology. This analysis focuses on collecting and analyzing data related to process performance and product quality to identify patterns and trends.

Consistently collecting data and setting clear objectives for data analysis are important. Involving all team members is crucial to ensure that objectives are met and to foster collaboration in implementing changes.

During data analysis, it is necessary to identify patterns and trends to assess process performance and product quality. Specific objectives should be established for each of these aspects, and changes should be implemented based on the results obtained.

In summary, data analysis is an essential tool for monitoring the effectiveness of industrial methodologies. Consistent data collection, clear objectives, and implementation of changes based on results are crucial.

Implementation of Changes

The implementation of changes is essential for continuous improvement in company processes and ensuring the effectiveness of industrial methodologies. Action plans should be established to implement changes based on the results obtained from industrial methodology review meetings, customer satisfaction surveys, and data analysis.

These action plans must be specific and detailed, including clear objectives, timelines, and responsibilities for each team member involved. Metrics should be established to evaluate the success of implemented changes, and constant monitoring is necessary to ensure that objectives are met.

Collaboration among team members in implementing changes should be encouraged, and opportunities for training and professional development should be provided to ensure that everyone is aligned and committed to the continuous improvement of processes.

In summary, the implementation of changes is essential for continuous improvement in company processes and ensuring the effectiveness of industrial methodologies. Specific and detailed action plans, metrics to evaluate success, and collaboration and professional development of team members are important.

Conclusion

The assessment and monitoring of the effectiveness of industrial methodologies are essential for continuous improvement in company processes, customer satisfaction, and process efficiency. Methodology review meetings, customer satisfaction surveys, data analysis, and the implementation of changes are essential tools for carrying out this assessment and monitoring process.

Clear objectives and specific action plans should be established for each of these tools, and collaboration and professional development of team members should be encouraged to ensure success in implementing changes and continuous improvement of processes.

In conclusion, the assessment and monitoring of the effectiveness of industrial methodologies are continuous processes that require constant commitment and a mindset of continuous improvement from all team members.

COMMUNICATION AND COLLABORATION IN THE IMPLEMENTATION OF INDUSTRIAL METHODOLOGIES

Implementation of industrial methodologies is a process that aims to improve efficiency and productivity in a company. These methodologies involve the application of specific tools and techniques to enhance the company's processes and systems. However, for the implementation of these methodologies to be successful, it is essential to have good communication and collaboration among different departments and teams within the company. This chapter explores the importance of communication and collaboration in the implementation of industrial methodologies.

Communication in the Implementation of Industrial Methodologies:

Communication is a key factor in the implementation of industrial methodologies as it enables different departments and teams within the company to work together to achieve common objectives. Effective communication allows teams to work more efficiently, identify and resolve issues more quickly, and make informed decisions.

In the context of implementing industrial methodologies, communication should be bidirectional. This means that it is not only important for managers and leaders to communicate with employees, but it is also crucial for employees to communicate with managers and leaders. Bidirectional communication allows employees to express their ideas, concerns, and questions, which can help improve the implementation of industrial methodologies.

Additionally, it is important for communication in the implementation of industrial methodologies to be clear and concise. Clear communication helps avoid misunderstandings and reduces the likelihood of errors. It is also important for communication to be in a language that is understandable to all team members, ensuring a common understanding of objectives and processes involved.

Collaboration in the Implementation of Industrial Methodologies:

Collaboration is another key factor in the implementation of industrial methodologies. Collaboration refers to the ability of different departments and teams within the company to work together to achieve common goals. In the context of implementing industrial methodologies, collaboration can help ensure that all team members are aligned with goals and work together to achieve them.

Effective collaboration in the implementation of industrial methodologies involves active participation from all team members. This means that all team members should be aware of their roles and responsibilities and work together to achieve common goals. Additionally, it is important to establish clear and effective communication channels to facilitate collaboration among team members.

Another important aspect of collaboration in the implementation of industrial methodologies is the ability of team members to share information and knowledge. This can help improve the quality of the company's processes and systems and ensure that all team members have a common understanding of the processes and objectives involved.

Benefits of Good Communication and Collaboration in the Implementation of Industrial Methodologies:

Good communication and collaboration in the implementation of industrial methodologies can generate several benefits for the company, including:

Improved Efficiency: When different departments and teams work together and share information, efficiency in the company can be improved. Team members can identify and resolve problems more quickly and efficiently, reducing the time taken to complete a task or project. Additionally, collaboration can help identify areas for improvement in processes, leading to greater efficiency in the future.

Error Reduction: Good communication and collaboration can also help reduce errors. By having a common understanding of processes and objectives, team members can avoid misunderstandings and make informed decisions, leading to a reduction in errors. This can save time and resources in error correction and rework.

Increased Productivity: The implementation of industrial methodologies aims to improve productivity in the company. Good communication and collaboration can help ensure that all team members are working together towards the same goals, thereby enhancing overall productivity. Additionally, when team members work together more efficiently, they can complete more tasks in less time, further improving productivity.

Improved Work Environment: Effective communication and collaboration can enhance the work environment in the company. By working together towards common goals, team members can develop stronger and more collaborative relationships, improving morale and job satisfaction. When employees feel valued and have a positive relationship with their colleagues, they are more likely to be more productive and willing to contribute to the company in the long run.

Quality Improvement: Good communication and collaboration can contribute to improving the quality of the company's processes and systems. By sharing information and knowledge, team members can identify areas for improvement and work together to enhance overall quality. Additionally, when employees are working together more efficiently and effectively, they can ensure that processes and systems are being followed correctly, improving the quality of the company's products and services.

Increased Innovation: Communication and collaboration can foster innovation in the company. By working together and sharing ideas, team members can generate new ideas and solutions to existing problems. Effective collaboration can help identify opportunities for improvement and evolution of existing processes and systems, leading to significant innovations in the company.

Enhanced Customer Satisfaction: Good communication and collaboration can contribute to improving customer satisfaction. By working together to enhance the quality of the company's products and services, team members can ensure that customers receive the best possible products and services. Additionally, when employees are working together more efficiently and effectively, they can ensure that orders are completed on time, further improving customer satisfaction.

Greater Adaptability: Good communication and collaboration can also help companies become more adaptable to changes in the market and industry. By working together and sharing information, team members can identify and respond more quickly to changes in market demand, new trends, and advances in technology. The ability to adapt quickly to these changes can be a competitive advantage for the company.

Improved Leadership: Good communication and collaboration can enhance leadership capabilities in the company. Leaders who encourage collaboration and effective communication can create a positive and collaborative work environment. Additionally, when leaders are involved in communication and collaboration, they can identify opportunities for improvement and lead change in the company.

Time and Resource Savings: Good communication and collaboration can help save time and resources in the company. By working together more efficiently, team members can complete tasks in less time, reducing labor costs. Additionally, by identifying and resolving problems more quickly and effectively, the cost of rework and error correction can be reduced.

Conclusion

In conclusion, communication and collaboration are crucial in the implementation of industrial methodologies. Effective communication allows different departments and teams within the company to work more efficiently, identify and resolve issues more quickly, and make informed decisions. Effective collaboration involves active participation from all team members and the ability to share information and knowledge. Good communication and collaboration can generate a range of benefits for the company, including improved efficiency, error reduction, increased productivity, improved work environment, quality improvement, increased innovation, enhanced customer satisfaction, greater adaptability, improved leadership, and time and resource savings. Therefore, it is important for companies to pay attention to communication and collaboration when implementing industrial methodologies and work to improve these aspects in their organizational culture.

IDENTIFICATION AND RESOLUTION OF PROBLEMS WITH INDUSTRIAL METHODOLOGIES

In industry, problems are an inevitable part of the production process. Often, these issues can impact the quality, efficiency, and profitability of a company. Therefore, it is essential to have an effective method for identifying and resolving problems in an industrial environment.

This chapter discusses industrial methodologies used to identify and solve problems in the industry. These methodologies include Six Sigma, Lean Manufacturing, and Total Quality Management (TQM). Additionally, common stages used in these methodologies will be discussed, along with some tools and techniques used for problem identification and resolution.

Six Sigma

Six Sigma is a methodology used to improve the quality of a company's products and services. It focuses on reducing variation in production processes and enhancing customer satisfaction. Six Sigma employs a data-driven approach to identify and resolve problems.

The Six Sigma methodology is based on a five-stage process known as DMAIC (Define, Measure, Analyze, Improve, and Control). Each stage has specific objectives and uses different tools and techniques to achieve these goals.

The first stage of DMAIC is Define, where the problem is defined, and project objectives are established. Customer and market needs are also identified.

The second stage is Measure, where data on the production process is collected to determine process capability and stability.

The third stage, Analyze, involves analyzing the data collected in the previous stage to identify the root causes of the problem. Statistical tools such as Pareto Analysis and Ishikawa Diagram are used to identify root causes.

The fourth stage, Improve, focuses on implementing solutions to resolve the problem. Tools such as Design of Experiments and Value Engineering are used to enhance the production process.

The final stage, Control, involves implementing control measures to ensure the production process remains stable and capable. Tools such as Control Plan and Control Chart are used to keep the process under control.

Lean Manufacturing

Lean Manufacturing is another methodology used to improve quality and efficiency in production. It concentrates on eliminating waste and enhancing the efficiency of the production process. Lean Manufacturing uses a continuous improvement approach to identify and solve problems.

The Lean Manufacturing methodology is based on a four-stage process known as the Plan-Do-Check-Act (PDCA) Cycle, which includes Planning, Doing, Checking, and Acting. Each stage has specific objectives and employs different tools and techniques to achieve these objectives.

The first stage of the PDCA cycle is Plan, where the problem is defined, and project objectives are established. Performance measures used to gauge project success are identified.

The second stage is Do, where solutions are implemented to solve the problem, using an action-oriented approach.

The third stage, Check, involves collecting data to assess the success of implemented solutions. Tools like Value Stream Mapping and Fishbone Diagram are used to evaluate process performance.

The final stage, Act, involves taking measures to continuously improve the process. Tools like Kaizen and A3 are used to enhance the production process and eliminate waste.

Total Quality Management (TQM)

Total Quality Management is a methodology used to improve quality and efficiency in production. It focuses on enhancing quality in all aspects of the company, including culture, processes, and products. TQM uses a continuous improvement approach to identify and solve problems.

The TQM methodology is based on an eight-stage process known as the Enhanced Deming Cycle (PDCA). The eight stages include Plan, Do, Check, Act, Analyze, Design, Implement, and Maintain. Each stage has specific objectives and uses different tools and techniques to achieve these objectives.

The first stage of the Enhanced Deming Cycle is Plan, where the problem is defined, and project objectives are established. Performance measures used to gauge project success are identified.

The second stage is Do, where solutions are implemented to solve the problem.

The third stage is Check, where data is collected to evaluate the success of implemented solutions.

The fourth stage is Act, where measures are taken to continuously improve the process.

The fifth stage is Analyze, where data collected in the previous stage is analyzed to identify the root causes of the problem.

The sixth stage is Design, where solutions are designed to solve the problem.

The seventh stage is Implement, where the designed solutions are implemented.

The last stage is Maintain, where control measures are implemented to ensure the stability and capability of the production process.

Tools and Techniques for Problem Identification and Resolution

There are several tools and techniques commonly used in industrial methodologies to identify and resolve problems. Some of these tools and techniques include:

Ishikawa Diagram: Also known as a Fishbone Diagram, it is used to identify the root causes of a problem.

Pareto Analysis: Used to identify the most significant problems affecting the quality and efficiency of the production process.

Control Chart: Used to monitor the production process and detect deviations from the process.

Control Plan: Used to establish control measures necessary to ensure the quality and efficiency of the production process.

Value Stream Mapping: Used to analyze the flow of the production process and identify areas of waste.

Kaizen: Refers to continuous improvement and is used to identify opportunities for improvement in the production process.

A3: A problem-solving tool used to document the improvement process and communicate results to stakeholders.

FMEA (Failure Mode and Effect Analysis): Used to identify potential failures in the production process and design control measures to prevent them.

5S: Refers to workplace organization, cleanliness, and standardization to improve the efficiency and safety of the production process.

Kanban: Used to manage inventory and ensure that necessary materials are available when needed.

These are just some of the tools and techniques used in industrial methodologies to identify and resolve problems. Each tool and technique has its own benefits and is used in different situations.

Conclusions

Industrial methodologies are a set of techniques and tools used in the industry to improve the quality and efficiency of the production process. These methodologies are based on the idea of continuous improvement and waste elimination, applied across a wide range of industrial sectors.

The PDCA cycle is one of the most widely used methodologies in the industry, focusing on the continuous improvement of the production process. This methodology is divided into four stages: Plan, Do, Check, and Act. Specific activities are carried out in each stage to identify and solve problems, and control measures are established to ensure the quality of the production process.

Another popular methodology is Total Quality Management, which focuses on continuous improvement in all aspects of the company. This methodology is based on the notion that quality is not only the responsibility of the quality control department but is the responsibility of all employees in the company. To implement Total Quality Management, specific activities such as employee training, waste identification and elimination, and the implementation of control measures to ensure production quality are carried out.

In addition to these methodologies, there are various specific tools and techniques used in the industry to identify and resolve problems. The Ishikawa Diagram is a tool used to identify the root causes of a problem, while Pareto Analysis is used to identify the most important problems that need to be addressed first. The Control Chart is used to monitor the quality of the production process and detect any unwanted variations.

In summary, industrial methodologies are essential to ensure the quality and efficiency of the production process. Specific tools and techniques used in these methodologies, such as the PDCA cycle, Total Quality Management, Ishikawa Diagram, and Pareto Analysis, allow for effective problem identification and resolution, as well as continuous monitoring and improvement of the production process. Managers and engineers should be familiar with these methodologies and tools to apply them effectively and achieve excellence in their production process. The implementation of industrial methodologies can help companies stay competitive and meet the needs and expectations of customers in an increasingly demanding market.

DEVELOPMENT OF ACTION PLANS AND IMPROVEMENT WITH INDUSTRIAL METHODOLOGIES

In any company or industry, the development of action and improvement plans is crucial to achieving sustainable growth and increasing process efficiency. Action and improvement plans consist of a set of strategies and actions implemented to enhance processes, optimize resources, and increase overall productivity.

This chapter will explain how industrial methodologies can be applied to develop effective action and improvement plans. It will also discuss the different stages of the improvement process and how they can be implemented in a company.

Industrial methodologies for developing action and improvement plans

Industrial methodologies comprise a set of techniques and tools used to optimize processes and enhance overall efficiency in a company. These methodologies have been developed over time and successfully applied across various industrial sectors.

Some of the most popular industrial methodologies include Lean Manufacturing, Six Sigma, Total Quality Management (TQM), Kaizen, and the Theory of Constraints (TOC).

Lean Manufacturing

Lean Manufacturing focuses on waste elimination and creating value for the customer. It is based on the principle that all activities not adding value to the customer should be eliminated.

Lean Manufacturing is divided into five stages: value identification, value stream mapping, flow creation, pull production system implementation, and continuous improvement.

The identification of value involves identifying the products or services that are valued by customers and eliminating those that are not. Value stream mapping is the process of visualizing the flow of materials and information through the production process. Creating flow involves eliminating bottlenecks and establishing a continuous production flow. The implementation of a pull production system involves producing products or services only when they are needed. Continuous improvement involves implementing measures to eliminate waste and improve the efficiency of the production process.

Six Sigma

Six Sigma aims to reduce variability in processes and improve product or service quality. It operates on the principle that process variability is the main cause of defects.

Six Sigma is divided into five stages: Define, Measure, Analyze, Improve, and Control (DMAIC). The Define stage involves defining the problem and identifying the project scope. The Measure stage involves collecting data and measuring the process. The Analyze stage involves identifying the root causes of the problem. The Improve stage involves implementing solutions to address the problem. The Control stage involves implementing measures to sustain the improvement.

Total Quality Management (TQM)

TQM focuses on continuous improvement of product or service quality and is based on the principle that quality is everyone's responsibility in the company.

TQM is divided into eight stages: leadership, education and training, employee involvement, customer focus, continuous improvement, measurement and results analysis, supplier management, and teamwork.

The leadership stage involves the commitment of top management to the implementation of continuous improvement. Education and training involve instructing employees in quality improvement techniques. Employee involvement entails the active participation of employees in continuous improvement. Customer focus involves meeting the needs and expectations of the customer. Continuous improvement entails implementing measures to continually enhance the quality of the product or service. Measurement and analysis of results involve monitoring and analyzing outcomes to take improvement actions. Supplier management involves the selection and evaluation of suppliers to ensure the quality of the product or service. Teamwork involves collaboration and cooperation among different departments within the company.

Kaizen

Kaizen concentrates on continuous process improvement and is based on the belief that all employees can contribute to continuous improvement.

Kaizen is divided into three stages: problem identification, problem analysis, and problem solution. Problem identification involves identifying areas that need improvement. Problem analysis involves identifying the root causes of the problem. Problem solution involves implementing measures to address the problem and prevent its recurrence.

Theory of Constraints (TOC)

TOC targets the identification and elimination of limitations in processes, operating on the principle that processes are constrained.

TOC is divided into three stages: identification of the constraint, exploitation of the constraint, and elevation of the constraint. Identifying the constraint involves recognizing the limitation in the process. Exploiting the constraint involves maximizing the use of that limitation. Elevation of the constraint involves eliminating or mitigating the limitation to improve the overall system performance.

Improvement Process Stage

The improvement process is divided into four stages: planning, implementation, monitoring, and evaluation.

Planning

The planning stage involves problem identification, definition of improvement objectives, identification of root causes of the problem, and selection of the improvement methodology.

Problem identification involves identifying areas that need improvement. Definition of improvement objectives involves defining desired outcomes. Identification of root causes of the problem involves identifying underlying causes of the problem. Selection of the improvement methodology involves choosing the most suitable methodology to solve the problem.

Implementation

The implementation stage involves putting into action the solutions identified in the planning stage.

This stage involves taking the necessary actions to improve the process and achieve the defined objectives. It also involves communicating and engaging employees in the implementation of solutions.

Monitoring

The monitoring stage involves tracking and measuring the results of the implemented solutions. This stage includes comparing the results with the objectives defined in the planning stage and identifying potential deviations.

Evaluation

The evaluation stage involves assessing the results and identifying additional improvement opportunities. This stage includes identifying factors that contributed to the success or failure of the implementation and learning lessons.

Continuous Improvement Tools

There are several tools and techniques that can be used for continuous process improvement. Some of the most common tools include:

Flowcharts: a tool used to visualize processes and the sequence of activities.

Pareto Charts: a tool used to identify the most significant problems and primary causes.

Ishikawa or Fishbone Diagrams: a tool used to identify the root causes of a problem.

Value Stream Analysis (VSA): a tool used to identify activities that add value to the process.

Process Maps: a tool used to visualize processes and interactions between different departments.

Cost-Benefit Analysis: a tool used to assess the costs and benefits of proposed solutions.

Conclusions

In conclusion, implementing action and improvement plans using industrial methodologies can help companies continuously improve their processes and products or services. These methodologies focus on problem identification, root cause identification, and solution implementation to improve processes. They also emphasize employee collaboration and commitment to continuous improvement.

It is important for companies to select the appropriate improvement methodology for their needs and commit to implementing the identified solutions. Additionally, it is crucial for companies to monitor and evaluate results to identify additional improvement opportunities and ensure that processes continue to improve continuously. Continuous improvement is an ongoing process that never ends, but it can help companies stay competitive in a changing market and meet the needs and expectations of their customers.

RISK AND OPPORTUNITY ASSESSMENT IN THE IMPLEMENTATION OF INDUSTRIAL METHODOLOGIES

The implementation of industrial methodologies is a common practice in operations management within companies. However, this implementation can involve risks and opportunities that must be assessed to ensure the success of the project. In this chapter, the evaluation of risks and opportunities in the implementation of industrial methodologies will be discussed. It will define what is meant by risk and opportunity, methodologies for risk and opportunity assessment will be explained, and the steps for conducting a proper assessment will be detailed. Additionally, some examples of common risks and opportunities that may arise in the implementation of industrial methodologies will be analyzed.

Definition of Risk and Opportunity:

Before delving into the assessment of risks and opportunities in the implementation of industrial methodologies, it is important to define what is meant by risk and opportunity. Risk is defined as the possibility of an event or situation occurring that negatively affects the project or organization. On the other hand, opportunity is defined as a situation that can be beneficial for the project or organization.

The proper evaluation of risks and opportunities in the implementation of industrial methodologies is crucial to avoid potential problems and maximize the project's benefits. A thorough assessment of risks and opportunities can provide valuable information for decision-making and planning the methodology's implementation.

Methodologies for Risk and Opportunity Assessment:

There are different methodologies for assessing risks and opportunities in the implementation of industrial methodologies. The most common ones are explained below:

SWOT Analysis

The SWOT analysis (Strengths, Weaknesses, Opportunities, Threats) is a strategic assessment tool that allows the identification of internal and external factors that may impact the success of a project. This analysis is divided into four categories:

Strengths: Internal factors that are positive for the project or organization.

Weaknesses: Internal factors that may negatively affect the project or organization.

Opportunities: External factors that can be beneficial for the project or organization.

Threats: External factors that may negatively impact the project or organization.

The SWOT analysis is useful for identifying the strengths and weaknesses of the project and the external opportunities and threats that may arise during the methodology implementation.

Risk and Opportunity Analysis

The risk and opportunity analysis is a methodology that identifies and evaluates potential events that can affect the project. The analysis is divided into two categories: risks and opportunities. Risks are events that can have a negative impact on the project or organization, while opportunities are events that can have a positive impact.

The risk and opportunity analysis is useful for identifying events that can negatively or positively affect the project and taking measures to reduce negative impacts and capitalize on opportunities.

Cost Analysis

Cost analysis is a methodology that evaluates the financial impact of implementing the methodology. The analysis focuses on the direct and indirect costs of the project and the potential gains and savings that can be achieved through methodology implementation.

Cost analysis is useful for assessing the financial viability of the project and determining whether the benefits outweigh the costs.

Steps for Risk and Opportunity Assessment

To carry out a proper assessment of risks and opportunities in the implementation of industrial methodologies, it is necessary to follow some key steps:

Identify potential risks and opportunities: In this stage, all possible risks and opportunities that may arise during the implementation of the methodology should be identified.

Evaluate the likelihood of risks and opportunities: In this stage, assess the likelihood of the identified risks and opportunities occurring. It is essential to classify them based on their probability of occurrence and potential impact on the project.

Identify possible mitigation measures: In this stage, identify possible measures to mitigate risks and capitalize on identified opportunities.

Evaluate the costs and benefits of mitigation measures: In this stage, evaluate the costs and benefits of the identified mitigation measures and determine their financial feasibility.

Implement mitigation measures: In this stage, implement the mitigation measures to reduce the negative impact of identified risks and capitalize on opportunities.

Examples of Risks and Opportunities in the Implementation of Industrial Methodologies

Below are some examples of common risks and opportunities that may arise in the implementation of industrial methodologies:

Risks:

Lack of adequate training: If personnel is not adequately trained in the methodology to be implemented, there may be implementation problems and an increase in project time and costs.

Resistance to change: If employees resist the changes involved in the implementation of the methodology, there may be delays and issues in implementation.

Technical problems: Technical problems may arise during the implementation of the methodology, leading to implementation delays and increased costs.

Opportunities:

Efficiency improvement: The implementation of an industrial methodology can improve the efficiency of organizational processes and reduce costs.

Quality improvement: The implementation of an industrial methodology can enhance the quality of products or services offered by the organization, increasing customer satisfaction and loyalty.

Increased productivity: The implementation of an industrial methodology can increase organizational productivity, allowing for higher production without increased costs.

Conclusions

Risk and opportunity assessment is a key step in the implementation of industrial methodologies. It allows for the identification of potential risks and opportunities, the evaluation of their potential impact, and the determination of necessary mitigation measures. Risk and opportunity assessment also enables the assessment of the financial viability of the project and the determination of whether the benefits outweigh the costs.

It is important to note that the implementation of an industrial methodology can have a significant impact on the organization. Therefore, conducting a proper assessment of risks and opportunities is crucial to ensure project success.

Additionally, it is important for the organization to provide adequate training to personnel and foster a culture of change and continuous improvement to ensure successful implementation of the methodology.

In summary, risk and opportunity assessment is a key step in the implementation of industrial methodologies. It allows for the identification of potential risks and opportunities and the determination of necessary mitigation measures to ensure project success. Proper assessment of risks and opportunities, along with adequate training for personnel, is essential for successful implementation of the methodology.

TRAINING AND EDUCATION IN INDUSTRIAL METHODOLOGIES

Industrial methodologies are techniques and tools used in the industry to improve the efficiency and quality of production processes. These methodologies include Lean Manufacturing, Six Sigma, Just in Time, Theory of Constraints, Total Productive Maintenance, among others.

Training and education in these methodologies are essential for companies to implement them effectively and achieve significant improvements in their production processes. This chapter will discuss the importance of training and education in industrial methodologies, the benefits they offer, and some examples of their application in different industry areas.

Importance of training and education in industrial methodologies

Training and education in industrial methodologies are fundamental to enhance the production processes of companies and increase their competitiveness in the market. The application of these methodologies allows for the optimization of production processes, cost reduction, improvement in product quality, and increased company efficiency.

The implementation of these methodologies requires trained personnel. Training and education in these methodologies enable staff to understand the importance of implementing these techniques and tools and how to apply them effectively in the production process.

Benefits of training and education in industrial methodologies

Training and education in industrial methodologies offer numerous benefits for companies, including the following:

Improved efficiency: The implementation of industrial methodologies allows for the identification of waste areas and process improvements, resulting in increased efficiency in the company.

Cost reduction: Identifying waste areas and improving production processes help reduce production costs.

Enhanced product quality: Implementing industrial methodologies identifies and corrects defect areas in the production process, leading to improved product quality.

Increased productivity: Implementing industrial methodologies optimizes production processes, translating into increased company productivity.

Improved customer service: Implementing industrial methodologies enhances efficiency and quality of customer service, resulting in increased customer satisfaction and loyalty.

Areas of application of industrial methodologies

Industrial methodologies can be applied in various industry areas, including:

Production processes: Industrial methodologies can be applied to improve efficiency, reduce production costs, and enhance product quality.

Maintenance: Industrial methodologies can be applied to machinery and equipment maintenance to reduce downtime and improve efficiency.

Distribution and storage: Industrial methodologies can be applied to improve efficiency in product distribution and storage, reduce delivery times, and enhance inventory management.

Project management: Industrial methodologies can be applied in project management to optimize resources, reduce delivery times, and improve project quality.

Supply chain management: Industrial methodologies can be applied in supply chain management to improve efficiency in supplier management, reduce costs, and enhance product quality.

Examples of application of industrial methodologies

Here are some examples of the application of industrial methodologies in different industry areas:

Lean Manufacturing: Focuses on eliminating waste in the production process. An example of its application would be reducing waiting times between operations in a production process, improving efficiency, and reducing production costs.

Six Sigma: Focuses on identifying and eliminating defects in the production process. An example of its application would be reducing the number of defects in a product, resulting in improved product quality.

Just in Time: Focuses on eliminating inventory and producing based on customer demand. An example of its application would be reducing waiting times in product delivery, improving customer service, and reducing inventory costs.

Theory of Constraints: Focuses on identifying and eliminating bottlenecks in the production process. An example of its application would be identifying a bottleneck in a production line and implementing measures to eliminate it, resulting in improved process efficiency.

Total Productive Maintenance: Focuses on improving machinery and equipment maintenance. An example of its application would be implementing a preventive maintenance program in an industrial plant, resulting in reduced downtime and improved plant efficiency.

Conclusion

Training and education in industrial methodologies are essential for companies to implement these techniques and tools effectively and achieve significant improvements in their production processes. Industrial methodologies offer numerous benefits, including improved efficiency, cost reduction, enhanced product quality, increased productivity, and improved customer service.

Industrial methodologies can be applied in different industry areas, such as production processes, maintenance, distribution and storage, project management, and supply chain management. The application of these methodologies in the industry can contribute significantly to improving the competitiveness of companies in the market.

CHANGE MANAGEMENT IN THE IMPLEMENTATION OF INDUSTRIAL METHODOLOGIES

Change management is a key aspect in the implementation of industrial methodologies, as it involves significant changes in processes, organizational culture, and interpersonal relationships. This chapter will address the main aspects related to change management in the implementation of industrial methodologies, including factors influencing implementation success, strategies for handling resistance to change, roles and responsibilities of change leaders, as well as tools and techniques available to facilitate implementation.

Factors influencing implementation success of industrial methodologies

The implementation of industrial methodologies is a complex process involving multiple factors influencing its success. In general, these factors can be classified into three main categories: organizational culture, change management capability, and planning and execution of implementation.

Organizational culture is one of the most important factors influencing the success of industrial methodology implementation. Organizational culture can be defined as the set of values, beliefs, norms, and behaviors that characterize an organization. A strong and cohesive organizational culture can be a determining factor in the success of industrial methodology implementation. Conversely, a weak or fragmented organizational culture can be a significant obstacle to successful implementation.

Change management capability is another critical factor influencing the success of industrial methodology implementation. Change management refers to the set of processes and activities carried out to plan, implement, and control changes in an organization. Good change management capability is essential to ensure that the implementation of the industrial methodology is carried out effectively and smoothly.

Planning and execution of implementation are also key factors influencing the success of industrial methodology implementation. Careful planning and efficient execution are necessary to ensure that the implementation is carried out effectively and that the desired objectives are achieved.

Strategies for handling resistance to change

Resistance to change is a common phenomenon in the implementation of industrial methodologies. Resistance to change can be a significant barrier to implementation success, as it can impede the change process and lead to employee dissatisfaction. The following are some strategies for handling resistance to change in the implementation of industrial methodologies:

Communicate the benefits of implementation: It is important for employees to understand the benefits of implementing the industrial methodology. Effective communication can help reduce resistance to change and motivate employees to support the implementation.

Involve employees: It is essential to involve employees in the implementation of the industrial methodology. This may include participation in workgroups, identification of problematic areas, and proposing solutions.

Provide training and support: Training and support are essential to help employees adapt to the changes occurring in the implementation of the industrial methodology. Training may include acquiring new skills and tools, while support may include guidance and follow-up to ensure that employees have the assistance they need for successful implementation.

Recognize and reward success: It is important to recognize and reward success in the implementation of the industrial methodology. This may include public recognition, financial incentives, and professional development opportunities. Recognition and rewards can motivate employees to support implementation and overcome resistance to change.

Accept and manage resistance: Finally, it is important to accept that resistance to change is a natural phenomenon in the implementation of industrial methodologies and must be managed effectively. Resistance management may include identifying underlying factors of resistance, taking measures to address these factors, and guiding employees through the change process.

Roles and responsibilities of change leaders

Change leaders play a fundamental role in the implementation of industrial methodologies. Change leaders are individuals who lead and coordinate the change process, ensuring that the implementation of the industrial methodology is carried out effectively and smoothly. The following are some key roles and responsibilities of change leaders in the implementation of industrial methodologies:

Establish a clear vision: Change leaders must establish a clear vision for the implementation of the industrial methodology. This may include defining clear objectives, communicating the vision to all employees, and aligning the vision with organizational values and culture.

Create a sense of urgency: Change leaders must create a sense of urgency around the implementation of the industrial methodology. This may include identifying risks and opportunities associated with non-implementation and communicating these risks and opportunities to all employees.

Develop and maintain a change coalition: Change leaders must develop and maintain a change coalition that supports the implementation of the industrial methodology. This may include identifying key stakeholders in the implementation, creating a project team, and collaborating with other leaders in the organization.

Communicate and educate: Change leaders must communicate and educate employees about the industrial methodology and the need for its implementation. This may include creating communication materials and organizing training and education sessions.

Implement and consolidate change: Finally, change leaders must implement and consolidate the change. This may include identifying obstacles and taking measures to overcome them, monitoring progress, and adapting the implementation as necessary.

Tools and techniques to facilitate the implementation of industrial methodologies

There are several tools and techniques that can be used to facilitate the implementation of industrial methodologies. These tools and techniques can help change leaders and employees plan, execute, and monitor the implementation process. The following are some of the most common tools and techniques used in the implementation of industrial methodologies:

Flowchart: A flowchart is a visual tool used to represent a process in terms of its individual steps and activities. Using a flowchart can help change leaders and employees visualize the implementation process and identify areas for improvement.

SWOT analysis: SWOT analysis is a tool used to assess the strengths, weaknesses, opportunities, and threats of an organization. Using SWOT analysis can help change leaders identify areas that require more attention during the implementation of the industrial methodology.

Gantt charts: Gantt charts are tools used to plan and monitor projects. Using Gantt charts can help change leaders and employees visualize the implementation schedule of the industrial methodology and identify any delays or deviations in the plan.

Responsibility matrix: The responsibility matrix is a tool used to identify the individual responsibilities of team members in a project. Using the responsibility matrix can help change leaders assign specific tasks and responsibilities to team members during the implementation of the industrial methodology.

Surveys and assessments: Surveys and assessments are tools used to collect information and feedback from employees about the implementation of the industrial methodology. Using surveys and assessments can help change leaders evaluate progress and identify any issues or challenges that need to be addressed.

Conclusion

The implementation of industrial methodologies can be a challenging and complex process. However, with proper planning, clear communication, and effective change management, implementation can be successful and can improve the efficiency, quality, and profitability of an organization. Change leaders play a crucial role in the implementation of industrial methodologies and must be willing to take responsibility for leading and coordinating the change process. The use of tools and techniques such as flowcharts, SWOT analysis, Gantt charts, responsibility matrices, surveys, and assessments can help facilitate the implementation of the industrial methodology and identify areas for improvement. In summary, the successful implementation of industrial methodologies requires careful planning, clear communication, effective change management, and collaboration between change leaders and employees.

INTEGRATION OF INDUSTRIAL METHODOLOGIES INTO THE COMPANY'S STRATEGY

The modern industry is characterized by fierce competition in both national and international markets, the increasing complexity of production processes, and the intensive use of advanced technologies. To remain competitive and profitable, companies need to adopt a comprehensive strategy that addresses all aspects of their operation, from planning and design to production and customer delivery. In this context, the integration of industrial methodologies is a key element for business success. This chapter explores the main industrial methodologies currently used and their impact on business strategy.

Industrial Methodologies

There are numerous industrial methodologies used today to improve the efficiency, quality, and profitability of production processes. Some of the most popular methodologies include:

Lean Manufacturing: This methodology focuses on waste elimination and continuous improvement of production processes. It is based on five fundamental principles: value, flow, pull, perfection, and respect for people. Implementing Lean Manufacturing involves identifying and eliminating activities that do not add value to the process, improving workflow, eliminating bottlenecks, and implementing on-demand production processes.

Six Sigma: This methodology focuses on improving quality and reducing variability in production processes. It relies on data collection and analysis to identify and solve problems, reduce defects, and enhance customer satisfaction. Implementing Six Sigma involves clearly defining objectives, collecting and analyzing data, implementing solutions, and continuously measuring results.

Theory of Constraints: This methodology focuses on identifying and eliminating constraints that limit a company's production capacity. It is based on the idea that a production chain is only as strong as its weakest link. Implementing the Theory of Constraints involves identifying bottlenecks and implementing solutions to improve production capacity at the weakest link.

Total Productive Maintenance (TPM): This methodology focuses on maximizing the efficiency of production equipment through preventive maintenance and continuous improvement. It is based on the idea that regular maintenance and early problem identification can prevent costly failures and extend the lifespan of equipment. Implementing TPM involves identifying critical equipment, defining preventive maintenance procedures, training workers, and implementing a system to monitor and measure equipment efficiency.

Integration of Industrial Methodologies into the Company's Strategy

The successful implementation of industrial methodologies requires a comprehensive business strategy that addresses all aspects of the company's operation. Some key elements of a comprehensive strategy include:

Clear definition of business objectives

A comprehensive strategy should start with a clear definition of business objectives. Objectives should be Specific, Measurable, Achievable, Relevant, and Timely (SMART). Additionally, objectives should align with the company's vision and mission. Once objectives are defined, key performance indicators can be established to measure progress toward achieving those objectives.

Identification of critical improvement areas

Identifying critical improvement areas is a crucial step in integrating industrial methodologies into the company's strategy. Critical improvement areas may include reducing production costs, improving quality, reducing delivery times, enhancing customer satisfaction, and improving operational efficiency. Once critical improvement areas are identified, the most suitable industrial methodologies can be selected to address those issues.

Implementation of industrial methodologies

Implementing industrial methodologies involves defining the processes and procedures necessary to apply those methodologies in the company's operation. This may include training workers, defining work procedures, implementing specific tools and technologies, and measuring performance. It is essential to ensure that all methodologies are integrated consistently into the company's daily operation.

Monitoring and performance measurement

Monitoring and measuring performance are key elements of a comprehensive strategy for integrating industrial methodologies. Performance indicators should be established to measure progress toward business objectives and continuous improvement. Results should be regularly monitored and evaluated to ensure that business objectives are being achieved and to identify additional areas for improvement.

Culture of continuous improvement

The integration of industrial methodologies into the company's strategy must be accompanied by a culture of continuous improvement. Workers should be encouraged to constantly seek ways to improve the company's processes and procedures. Regular feedback and active worker participation are key elements in fostering a culture of continuous improvement.

Conclusion

The integration of industrial methodologies into the company's strategy is essential to remain competitive in the modern industry. Industrial methodologies can improve the efficiency, quality, and profitability of production processes and can be used to address a wide range of business problems. Successful implementation of industrial methodologies requires a comprehensive strategy that addresses all aspects of the company's operation, from defining objectives to fostering a culture of continuous improvement. By consistently integrating industrial methodologies into the company's daily operation, businesses can enhance their competitiveness and profitability in both national and international markets.

EXPERIENCES AND SUCCESS CASES IN THE IMPLEMENTATION OF INDUSTRIAL METHODOLOGIES

In the world of industry, the implementation of methodologies is a common practice aimed at improving the efficiency, productivity, and quality of production processes. However, carrying out this task is not always easy and can pose various challenges. In this chapter, we will explore some experiences and success cases in the implementation of industrial methodologies with the aim of better understanding the challenges and benefits of this process.

Experiences in the implementation of industrial methodologies:

To begin with, it is important to note that there is no one-size-fits-all methodology that works for all companies and situations. Each organization has its own needs, goals, and resources, so it is important to adapt the methodology to its specific context. Below are some experiences and common challenges in the implementation of industrial methodologies.

Experience 1: Lean Manufacturing Implementation

A metal parts manufacturing company decided to implement Lean Manufacturing methodology to improve the efficiency and productivity of its processes. The implementation took place in several phases, starting with the identification of critical processes and the elimination of waste. The Kanban system was also implemented to improve inventory management and reduce lead times.

One of the main challenges was employee resistance to change. Initially, many were reluctant to abandon their old ways of working and adapt to the new processes. However, through training and active involvement of workers in the implementation process, this barrier was overcome.

Another significant challenge was the measurement and tracking of results. Although a significant improvement in efficiency and productivity was observed, establishing a baseline and measuring the exact impact of Lean Manufacturing implementation on final results proved challenging. However, the company continued to make adjustments and improvements in the process, resulting in continuous improvement in outcomes.

Experience 2: Six Sigma Implementation

A telecommunications company decided to implement the Six Sigma methodology to improve the quality of its processes and reduce defects in its products. A dedicated team was formed to implement Six Sigma, responsible for identifying critical processes and analyzing data to identify the causes of problems.

One of the most important challenges was the lack of understanding of the methodology by employees. Many did not understand how Six Sigma worked and how they could contribute to the improvement process. To overcome this obstacle, a training and communication campaign was conducted to explain the methodology and its benefits.

Another challenge was the collection and analysis of data. The company found that it lacked the necessary systems to effectively collect data, so it had to invest in technology and data analysis tools. Once this issue was resolved, there was a significant improvement in product quality.

In addition to the mentioned challenges, the implementation of Six Sigma also brought significant benefits to the company. For example, the methodology helped the company standardize its processes, reducing variability and improving product consistency. It also identified hidden problems and addressed them before they affected product quality.

Another significant benefit was the improvement in customer satisfaction. By reducing defects in products, the company was able to deliver higher-quality products to its customers, improving their satisfaction and loyalty. This, in turn, had a positive impact on the company's reputation and financial success.

Experience 3: Total Productive Maintenance (TPM) Implementation

A food manufacturing company decided to implement TPM methodology to improve the efficiency and reliability of its production equipment. The implementation was divided into several phases, starting with the identification of critical equipment and conducting a failure analysis to determine the root causes of problems.

One of the most significant challenges was the lack of employee commitment to the methodology. Initially, many workers did not see the value in dedicating time and resources to TPM implementation and preferred to stick to their old ways of working. To overcome this obstacle, the company conducted a communication and training campaign to explain the methodology and its benefits to employees.

Another significant challenge was the lack of resources for TPM implementation. The company found that it did not have enough trained personnel to effectively carry out the implementation, so it had to invest in training and hire new employees to meet the needs.

Despite these challenges, TPM implementation had significant results for the company. Downtime of equipment was significantly reduced, and its reliability improved. Additionally, the efficiency of maintenance processes was improved, reducing costs and improving customer satisfaction by reducing delivery times.

Success Cases in the Implementation of Industrial Methodologies:

In addition to these experiences, there are several success cases in the implementation of industrial methodologies in different companies and sectors. Below are some of them:

Case 1: Toyota and the Implementation of Lean Manufacturing

Toyota is one of the most well-known examples of success in the implementation of Lean Manufacturing. The company has been a pioneer in the development and implementation of this methodology since the 1950s, significantly contributing to its success as an automobile manufacturer.

The implementation of Lean Manufacturing at Toyota is based on the following principles:

Waste elimination: The company focuses on identifying and eliminating any activity that does not add value to the production process.

Continuous improvement: Toyota constantly seeks to improve its processes and products, using customer and employee feedback to identify improvement opportunities.

Teamwork: The company encourages collaboration and teamwork among employees from different departments and hierarchical levels.

Just in time: Toyota's Lean Manufacturing methodology is based on the concept of just-in-time production, meaning that products are manufactured in the exact quantity and timing needed, without inventory accumulation.

Total quality: Toyota focuses on the total quality of its products, involving all employees in problem prevention and the identification of improvement opportunities.

Thanks to the implementation of Lean Manufacturing, Toyota has significantly reduced delivery times, improved the quality of its products, and reduced production costs. Additionally, the methodology has allowed the company to adapt quickly to market changes and maintain its position as a leader in the automotive industry.

Case 2: Johnson & Johnson and the Implementation of Six Sigma

Johnson & Johnson, a leading company in the health sector, decided to implement the Six Sigma methodology to enhance the quality and efficiency of its production processes. The company focused on implementing Six Sigma in its manufacturing processes, including the identification of critical processes, employee training, and the implementation of continuous improvement tools.

As a result of implementing Six Sigma, Johnson & Johnson significantly reduced defects in its products, improved process efficiency, and lowered production costs. Moreover, the methodology enabled the company to enhance customer satisfaction by delivering higher-quality products and reducing delivery times.

Case 3: GE and the Implementation of Total Quality Management

General Electric (GE) is a company recognized for its successful implementation of Total Quality Management (TQM). The TQM implementation at GE focused on continuous process improvement, waste elimination, and the involvement of all employees in problem prevention and the identification of improvement opportunities.

The implementation of TQM at GE had a significant impact on the company's efficiency and quality. For example, the company managed to significantly reduce product delivery times, improve product quality, and reduce production costs. Furthermore, the methodology allowed GE to enhance customer satisfaction and maintain its position as an industry leader.

Conclusion

The implementation of industrial methodologies can pose significant challenges, such as resistance to change and a lack of resources. However, it can also yield substantial benefits, including improved process and product efficiency, cost reduction, and enhanced customer satisfaction.

Through the experiences and success cases presented in this chapter, it is evident that the implementation of industrial methodologies can be a powerful tool for improving competitiveness and financial success across different sectors. To achieve successful implementation, strong commitment and effective communication with employees, along with investment in training and resources, are crucial to ensuring proper implementation and the continuation of continuous improvement.

Additionally, it is essential to recognize that there is no universally applicable methodology; each company must find the one that best suits its needs and characteristics. Careful evaluation of different options and adaptation to the company's culture and strategy are important considerations.

Finally, it is crucial to view the implementation of industrial methodologies not as an isolated project but as a continuous and evolving process. Continuous improvement should be ingrained in the company's culture and long-term strategy.

In summary, the implementation of industrial methodologies can be a powerful tool for improving competitiveness and financial success for companies. However, it is vital for each company to find the methodology that best suits its needs, and the implementation should be viewed as a continuous and evolving process.

CHALLENGES AND OPPORTUNITIES IN THE FUTURE OF INDUSTRIAL METHODOLOGIES

Industrial methodologies have experienced rapid advancement in recent decades, driven by the increasing demand for improvements in efficiency and quality in production processes. These methodologies, ranging from Six Sigma and Lean Manufacturing to Lean Production and Total Quality Management, have proven to be extremely effective in helping companies enhance their processes and increase profitability.

However, today, businesses face a series of increasingly complex challenges in the business environment. Globalization, digitization, and automation are changing the industrial landscape, creating new opportunities and challenges. This chapter will analyze the challenges and opportunities facing industrial methodologies in the future, presenting some innovative solutions to address them.

Challenges of Industrial Methodologies

Adaptation to the Digital Era

Digitization has transformed the way industrial processes are carried out, creating new opportunities for efficiency and optimization. However, many companies have not yet adapted to this new era. Incorporating new technologies, such as the Internet of Things (IoT) and artificial intelligence (AI), requires a significant investment in infrastructure and human resources. The lack of training and expertise can be an obstacle to the effective implementation of these technologies.

Flexibility in the Supply Chain

Globalization has allowed companies to access new markets and suppliers, but it has also created new complexities in the supply chain. Companies need to be able to adapt quickly to changes in demand and market conditions, requiring increased flexibility in the supply chain. Additionally, companies need to be prepared for inherent risks in the supply chain, such as natural disasters, political or social disruptions, and cyber-attacks.

Talent Management

The success of any industrial methodology largely depends on the talent and experience of the people implementing it. However, talent shortages in some sectors and regions can hinder the effective implementation of industrial methodologies. Moreover, the aging workforce and a lack of relevant skills in the younger workforce can limit companies' ability to implement new technologies and methodologies.

Sustainability

Companies are increasingly concerned about the environmental and social impact of their operations. Industrial methodologies can help companies reduce their environmental impact and improve sustainability. However, it is also essential to consider the long-term effects of processes and products. Sustainability also includes social considerations, such as working conditions and human rights throughout the supply chain.

Opportunities for Industrial Methodologies

Automation

Automation of industrial processes can provide increased efficiency and a reduction in human errors. Robotics and automation can help companies adapt to the growing demand for customization and variability in production, enhancing efficiency and reducing costs. Automation can also be a solution to talent shortages in some sectors, allowing companies to do more with fewer personnel.

Data Analytics

Data analytics can be a valuable tool for companies looking to improve efficiency and quality in production processes. Real-time data collection and analysis can provide insights into equipment performance, product quality, and process efficiency. This data can be used to identify areas for improvement and make informed decisions about the implementation of industrial methodologies.

Integrated Management Systems

Integrated management systems can help companies effectively manage production processes and the supply chain. These systems allow centralized management of processes, reducing complexity and improving efficiency. Additionally, integrated management systems can enhance communication between departments and suppliers, increasing transparency and reducing errors.

Sustainability

Sustainability can be an opportunity for companies looking to differentiate themselves in the market and meet the demands of increasingly environmentally conscious consumers. The implementation of industrial methodologies can help companies reduce their environmental impact and improve the sustainability of their operations. This may include waste and emission reduction, the use of renewable energy, and improved energy efficiency.

Innovative Solutions to Address Challenges

Training and Skill Development

Training and skill development can help companies address talent shortages and the lack of relevant skills in the workforce. Companies can invest in training programs to update workers' skills and prepare them for the implementation of new technologies and methodologies. Additionally, companies can collaborate with educational and governmental institutions to promote the development of skills relevant to the industry.

Staggered Implementation of Technologies

Staggered implementation of technologies can help companies tackle the challenge of adapting to the digital era. Instead of attempting to implement all technologies at once, companies can gradually implement technologies and assess their impact on processes. This approach can help companies reduce risk and initial investment, allowing them to adjust their strategies as they learn more about the technologies.

Supply Chain Collaboration

Collaboration in the supply chain can help companies improve the flexibility and resilience of their operations. Companies can collaborate with suppliers and customers to share information and plan jointly to reduce the risk of supply chain disruptions. Additionally, collaboration can help companies identify opportunities to improve efficiency and reduce costs throughout the supply chain.

Open Innovation

Open innovation can help companies develop innovative solutions to address current and future challenges. Open innovation involves collaborating with external partners, such as startups, universities, and other stakeholders in the business ecosystem, to develop new solutions and technologies. Open innovation can allow companies to access new ideas and skills, helping them stay at the forefront of innovation in their industry.

Use of Emerging Technologies

The use of emerging technologies, such as artificial intelligence, robotics, and augmented reality, can provide new solutions for current and future industrial challenges. These technologies can improve efficiency and accuracy in production, reduce downtime, and enhance worker safety. Additionally, these technologies can enable greater customization and variability in production, improving customer satisfaction and competitiveness.

Focus on Quality

A focus on quality can help companies improve customer satisfaction and efficiency in production processes. Implementing quality management systems, such as ISO 9001, can help companies establish clear and standardized processes to ensure the quality of products and services. Moreover, a focus on quality can help companies identify areas for improvement and reduce non-quality costs.

Conclusions

The future of industrial methodologies presents challenges and opportunities for companies. Adapting to the digital era, talent shortages, and the demand for sustainability are some of the challenges companies face. However, innovative solutions, such as training and skill development, staggered implementation of technologies, supply chain collaboration, open innovation, the use of emerging technologies, and a focus on quality, can help companies address these challenges.

Companies that can adapt to these challenges and leverage these opportunities will be better positioned to compete in an increasingly demanding and changing business environment. The implementation of industrial methodologies can help companies improve efficiency, quality, and sustainability in their operations, enhancing customer satisfaction and reducing costs. Furthermore, the implementation of industrial methodologies can help companies stay at the forefront of innovation in their industry and ensure long-term success.

CONCLUSIONS AND RECOMMENDATIONS FOR THE IMPLEMENTATION OF INDUSTRIAL METHODOLOGIES

The implementation of industrial methodologies is a key topic in the efficient management of companies. These methodologies have been developed over the years with the aim of optimizing production processes, reducing costs, and improving product quality. In this chapter, the main conclusions and recommendations for the implementation of industrial methodologies will be presented.

The implementation of industrial methodologies is essential for improving the efficiency of production processes in companies. These methodologies help reduce production times, enhance product quality, and lower costs.

One widely used methodology in the industry is Lean Manufacturing. This methodology focuses on eliminating waste in production processes and continuous improvement.

Another widely adopted methodology is Six Sigma, which aims to reduce variability in production processes. It identifies and corrects issues in processes, leading to improved product quality.

The Cleaner Production methodology is one of the newest in the industry, focusing on reducing the environmental impact of production processes through the implementation of cleaner and more sustainable practices.

Implementing industrial methodologies requires a systemic approach and active involvement of workers. It is crucial to engage all levels of the organization in the implementation process and promote a culture of continuous improvement.

Worker training and education are essential for implementing industrial methodologies. Employees need training in specific methodologies and continuous improvement of production processes.

Change management is a critical factor in implementing industrial methodologies. Effectively communicating the changes being made and ensuring workers understand and accept these changes are important.

Measuring and monitoring performance indicators are essential to evaluate the effectiveness of implemented industrial methodologies. Performance indicators must be clearly defined and aligned with the company's objectives.

Feedback is key to continuous improvement in production processes. Workers should be able to provide feedback on processes, and this feedback should be considered in the implementation of improvements.

Recommendations

Before implementing an industrial methodology, it is important to assess the specific needs of the company and determine the most suitable methodology to meet these needs.

Implementation of industrial methodologies should be led by a multidisciplinary team, including representatives from all levels of the organization, committed to continuous improvement.

Worker training and education are essential for implementing industrial methodologies. Workers must receive proper training to effectively implement and use methodologies. Training should be continuous and include both theory and practical implementation of methodologies.

Involving workers in the implementation of industrial methodologies is important. This can be achieved through effective communication of objectives and benefits, as well as participation in continuous improvement teams.

Change management is crucial for the success of industrial methodology implementation. Establishing a communication and change management plan to involve workers and facilitate the transition to new practices is important.

Defining performance indicators before implementing industrial methodologies is advisable. These indicators should be relevant, measurable, and aligned with the company's objectives. Additionally, establishing a plan for monitoring and evaluating performance indicators is important.

Feedback is a key component for continuous improvement in production processes. Establishing feedback channels for workers to provide comments and suggestions on production processes is important. This feedback should be considered in the implementation of improvements.

Implementing industrial methodologies is not a one-time process. It is important to establish a continuous improvement plan that allows the evolution of practices and the incorporation of new methodologies or practices for greater efficiency and quality in production processes.

Establishing strategic alliances with suppliers and customers to implement sustainable and continuous improvement practices throughout the supply chain is advisable.

Conclusion

The implementation of industrial methodologies is a key tool for improving the efficiency and quality of production processes in companies. Lean Manufacturing, Six Sigma, and Cleaner Production are some of the most widely used methodologies in the industry.

The implementation of these methodologies requires a systemic approach, active involvement of workers, continuous training and education, change management, definition of performance indicators, and constant feedback.

Companies should evaluate their specific needs and determine the most suitable methodology. It is essential to establish a continuous improvement plan for the evolution of practices and the incorporation of new methodologies for greater efficiency and quality in production processes.

In summary, the implementation of industrial methodologies is a continuous and dynamic process that requires commitment, collaboration, and active participation at all levels of the organization. Effective implementation of these methodologies can lead to significant improvements in the efficiency, quality, and sustainability of company production processe.

ABOUT THE AUTHOR

Industrial and Systems Engineer

Master in Administration with Quality and Productivity

3 Certifications

2 Technical Studies

More than 20 training courses

Winner of the Singapore Cooperation Programme ITE

Instructor, engineer, content creator, and writer.
Discover the modern industry with the most controversial engineer.
engr's Workshop

I. Laisequilla

Author / Engineer